Japan's Transnational Environmental Policies

Labour, Education & Society

Edited by György Széll,
Heinz Sünker, Anne Inga Hilsen
and Francesco Garibaldo

Volume 25

PETER LANG
Frankfurt am Main · Berlin · Bern · Bruxelles · New York · Oxford · Wien

Rüdiger Kühr

Japan's Transnational Environmental Policies

The Case of Environmental
Technology Transfer to Newly
Industrializing Countries

PETER LANG
Internationaler Verlag der Wissenschaften

Bibliographic Information published by the Deutsche Nationalbibliothek
The Deutsche Nationalbibliothek lists this publication in the Deutsche Nationalbibliografie; detailed bibliographic data is available in the internet at http://dnb.d-nb.de.

Zugl.: Osnabrück, Univ., Diss., 2011

D 700
ISSN 1861-647X
ISBN 978-3-631-62089-2
© Peter Lang GmbH
Internationaler Verlag der Wissenschaften
Frankfurt am Main 2011
All rights reserved.

All parts of this publication are protected by copyright. Any utilisation outside the strict limits of the copyright law, without the permission of the publisher, is forbidden and liable to prosecution. This applies in particular to reproductions, translations, microfilming, and storage and processing in electronic retrieval systems.

www.peterlang.de

Foreword

塵も積もれば山となる[1]

This study on Japan's transnational environmental policy through the transfer of environmental technologies into Newly Industrialized Countries (NICs) grew out of a large number of small experiences throughout my life until now.

Growing up in the relatively freedom of my home town, Attendorn, surrounded by the small mountains, forests and lakes of Germany's Sauerland, I have always been closely attached to the environment, making me a pupil's activist in environmental protection measures, but also getting involved in the environmental politics of the city council. Today, now living in Hamburg, I enjoy every morning brining my eldest son Louis into his kindergarten in the forest and hence starting the day with the typical sounds, smells and optical impressions of nature.

The long-lasting and deep correspondence with my former pen friend Shizu *Onodera* did raise my deep interest in Japan, finally also supporting my decision to make Japan a key-focus during my MA studies in political science, geography and psychology in Münster followed by my PhD studies in political and social sciences at the Japan Research Centre in Osnabrück under Prof. György *Széll*.

Thanks to Dr. Juha *Uitto* and Prof. Tomonaga *Tairako* and the generous support of the Schneyder-Sasakawa-Fund, the Doctoral Scholarship of the German State of Lower Saxony, and the German Academic Exchange Service (DAAD) Scholarship Fund, I was provided with numerous opportunities between 1995 and 1999 to stay, work and research at the United Nations University (UNU) and Hitotsubashi University in Tokyo. These were times during which the discussions around sustainable development were becoming more concrete, bringing me also in close contact with Prof. Udo-Ernst *Simonis* at the Berlin-based Social Science Research Centre (WZB).

During my search on how to make sustainable development a practical opportunity for societies and which strategic approaches appear promising, technological solutions moved more and more into the foreground, though I know today on the one side how indispensable they can be, but on the other side what environmental challenges the production, usage and final disposal of (environ-

---

[1] Chiri mo tsumoreba yama to naru = Every piece of dust one day becomes a mountain (Japanese saying). It also means "Little and often fills the purpose" or "Many a little makes a mickle" = Many small amounts accumulate to make a large amount.

mental) techniques can also bring along. Similar to parents, teachers and professors I had the hope that nations, as the legitimate actors in the international community, would also attempt to pass on their experiences to those striving for the same level of development and joy. This moved transnational environmental policies through environmental technology transfer and eco-political development aid into my research focus, which has expanded since but still remains centred on these questions.

To my large pleasure both Prof. Széll and Prof. Simonis agreed to supervise my research and were unrelenting with their never-ending patience, gentle reminders but above all, with their trust that I would one day successfully and satisfactorily complete my writing, though my research might never be over.

Many more have contributed to this study but Prof. Iwao *Kobori*✝ deserves special reference, as his large network and kind support was a door-opener in Japan. I also learned a lot from accompanying Prof. Jun *Ui*✝, a well-known environmental scientist and key-person in Japan's environmental movements. His lecture notes were used as guidelines by organizers of anti-pollution movements and had a considerable impact on various other citizens' movements. During our joint missions on behalf of UNU, Prof. Motoyuki *Suzuki* provided me with further insights into environmental policies and strategies towards sustainability.

UNU's librarians Mayako *Matsuki* and Chiyoko *Hayashi*✝ helped me to get access to documents, something which may have been a rather challenge at a typical university. Holger *Rindermann* and Wesley *Crock* deserve special thanks for their critical reviews of my manuscripts and kind native-speaker editing.

Special thanks also to the 42 experts I had a chance to discuss the topics of this study with.

My family – Dr. Julia *Kastrup* and our sons Louis and Titus – but also my parents Gerda and Gerhard *Kühr* might not know the details where and when they contributed and inspired me. But they did by allowing me spend countless late nights in the office and re-showing me the miracles of life and earth with fresh eyes.

All these contributions accumulate to make this study finally possible, much in the same way that every single contribution towards sustainability is important in one day reaching this real utopian aim.

Rüdiger Kühr

Hamburg, February 2011

Contents

Figures & Tables xi
Abbreviations xiii

Part A – Approach 1
1 Introduction 1
2 Global Changes: Approaching the Research Object 7
   2.1 Paradigm – Sustainable Development 7
   2.2 Transnationalization – The Challenge 14
3 Definitions – The Key Terms 22
   3.1 Environment 22
   3.2 Technology 25
   3.3 Environmental Technology 26
      3.3.1 General Understanding 26
      3.3.2 International Attempts of Definition 27
      3.3.3 Environmental Technology Transfer Institutions 30
      3.3.4 Categorization of Environmental Technology 31
   3.4 Newly Industrializing Countries 34
      3.4.1 Rapid Economic Growth and the Environment 36
4 Environmental Problems - A Challenge for Nation-States 41
   4.1 Categorization of Environmental Problems 41
   4.2 Transnational Environmental Problems 42
5 Focus of the Study 46
6 Theoretical and Methodological Approximations 50
   6.1 Theoretical Framework 50
   6.2 Changing Role of States 50
   6.3 Academic Discussion 52
      6.3.1 Neo-Realists and Globalists 53
      6.3.2 Specific Approaches 56
7 Methodical Approach 65
   7.1 Actors Analysis 66
   7.2 Policy-Cycle 69
   7.3 Comparison 73
   7.4 Empirical Research 74
   7.5 Qualitative and Quantitative Methods 76

Part B – Background 77
8 History of Transnational Environmental Cooperation 77
   8.1 Multilateral 78

    8.1.1   Case of the "Whaling Convention" 81
    8.1.2   Case of the "Kyoto Protocol" 84
  8.2   Bilateral 91
9   Framework Conditions 94
  9.1   Impact on the Environment 94
    9.1.1   Imports of Forest & Agriculture Products 94
    9.1.2   Imports of Resources 97
        9.1.2.1   Case of Aluminium 98
    9.1.3   Greenhouse Gas Emissions 99

Part C – Analysis 103
10   Initiation 103
  10.1   Development of Transnational Relations 103
  10.2   Phase of Ecological Ignorance 106
  10.3   Growing Environmental Consciousness 111
  10.4   Growing Public Development Assistance 114
    10.4.1 Quantification of Official Development Assistance (ODA) 116
        10.4.1.1 Growing Environmental Orientation in ODA 123
  10.5   Political Motives of Initiation 125
11   Estimation 128
  11.1   International Conferences 128
  11.2   Institutional Changes in Japan's ODA 133
    11.2.1 ODA Charta 135
12   Selection 138
  12.1   Basic Environment Law and Basic Environment Plan 138
  12.2   Green Aid Plan 139
    12.2.1 Green Aid Plan for Indonesia 141
  12.3   National Action Plan for "Agenda 21" 144
13   Implementation 147
  13.1   Institutional Framework 147
  13.2   Consulting Companies 147
  13.3   Environmental Technology Transfer by Local Organizations 150
    13.3.1 Kitakyushu International Techno-Cooperative Association 150
    13.3.2 International Environmental Technology Centre 151
    13.3.3 International Centre for Environmental Technology Transfer 152
        13.3.3.1 Palembang Eco-Phoenix Plan Project 154
    13.3.4 Centre for International Transfer of Environmental Techniques ... 155
  13.4   Environmental Technology Transfer by National Agencies 157
    13.4.1 Japan International Cooperation Agency Projects 157
        13.4.1.1 Integrated Air Quality Management 157

    13.4.1.2 Environmental Management Centre 158
   13.4.2 Overseas Economic Cooperation Fund Projects 160
  13.5   Environmental Technology Transfer by Private Industry 161
  13.6   Environmental Technology Transfer by Non-Governmental
           Organizations 163
14   Evaluation 165
  14.1   Criteria of Effects 166
   14.1.1 Depth of Effect 166
   14.1.2 Breadth of Effect 166
   14.1.3 Speed of Effect 167
   14.1.4 Exactness of Effect 167
  14.2   Criteria of Effects in Context 167
  14.3   Hierarchy of Motives 170
15   Termination 175

Part D – Conclusions 177
16   Conclusions 177

17   Sources 185
  17.1   Bibliography 185
  17.2   Internet 214
  17.3   Press & Radio 215
  17.4   Experts 216

18   Appendix 219
  18.1   Index of Japanese terms 219
  18.2   Index of German terms 220
  18.3   Japan - International environmental treaties in force 221
  18.4   Germany - International environmental treaties in force 224

Figures & Tables

a) Figures

Figure 1: Japanese and German expressions of sustainable development 12
Figure 2: Population Density & Internet Distribution (2009) 17
Figure 3: Categorization of Environmental Technology 32
Figure 4: Simplified Model of Actors' Actions 68
Figure 5: Dynamic, open Policy-Cycle Model 71

b) Tables

Table 1: Geographic distribution of Internet Protocol Locations
(2007) 16
Table 2: ODA net performance
(2008, 1998 & 1994) 47
Table 3: Gross Domestic Product – Japan, USA, Germany
(2002 & 2008) 94
Table 4: Timber Imports – Japan, USA, Germany
(1998) 95
Table 5: Trade in Forest Industry Products – Japan, USA, Germany
(2002) 96
Table 6: Imports of Selected Raw Materials – Japan & Germany
(1997 & 1998) 98
Table 7: Emissions of Greenhouse Gases, NO_X and SO_X – Japan, USA,
Germany (2002) 100
Table 8: Bilateral Grant Aid – Japan & Germany
(1994, 1998, 2008) 119
Table 9: Bilateral Technical Cooperation – Japan & Germany
(1994, 1998, 2008) 120
Table 10: Total Bilateral Development Aid – Japan & Germany
(1994, 1998, 2008) 121
Table 11: Multilateral Development Aid – Japan & Germany
(1994, 1998, 2008) 122
Table 12: Japan's Environmental Cooperation
(1989-1998) 128

Abbreviations

ADB	Asian Development Bank
AHK	Außenhandelskammer, Germany
AiF	German Federation of Industrial Research Associations
AOTS	Association of Overseas Technical Scholarship, Japan
ASEAN	Association of South-East Asian Nations
BAPEDAL	Environmental Impact Management Agency, Indonesia
BEL	Basic Environment Law, Japan
BEP	Basic Environment Plan, Japan
BfAI	Bundesstelle für Außenhandelsinformationen, Germany
BGB	Bundesgesetzblatt, Germany
BMU	German Federal Ministry for the Environment, Nature Conservation and Nuclear Safety
BMZ	German Federal Ministry for Economic Cooperation and Development
BSE	Bovine Spongiform Encephalopathy
BUND	Friends of the Earth Germany
bUS$	Billion US-Dollar
CDG	Carl-Duisberg Society, Germany
CER	Certified Emission Reduction
CGIAR	Consultative Group on International Agricultural Research
CITES	Convention on International Trade in Endangered Species of Wild Fauna and Flora
CO_2	Carbon Dioxide
COP	Conference of Parties
CSD	Commission on Sustainable Development
DAAD	German Academic Exchange Service
DAC	Development Assistance Committee
DBU	German Environment Foundation
DED	German Development Service
DEG	Deutsche Investitions- und Entwicklungsgesellschaft, Germany
DIHT	German Chamber of Industry and Commerce
DSE	German Foundation for International Development
EA	Environmental Agency, Japan
ECFA	Engineering Consulting Firms Association
ECOSOC	United Nations Economic and Social Council
ENTRI	Environmental Treaties and Resource Indicators
EPA	Economic Planning Agency
ET	Environmental Technology

ETC	Strengthening Environmental Technological Capability in Developing Countries
EST	Environmentally Sound Technology
EU	European Union
FRG	Federal Republic of Germany
FZ	Financial Cooperation
G7	Group of seven nations (Canada, France, Germany, Italy, Japan, the United Kingdom, and the United States)
GDP	Gross Domestic Product
GDR	German Democratic Republic
GEC	Global Environment Centre Foundation, Japan
GIZ	German Society for International Cooperation
GJCC	German Japanese Cooperation Council for High Tech and Environmental Technology
GTZ	Deutsche Gesellschaft für Technische Zusammenarbeit, Germany
ICETT	International Center for Environmental Technology Transfer, Japan
ILEC	International Lake Environment Committee Foundation; Japan
IETC	International Environmental Technology Centre
IMF	International Monetary Fund
INFID	International NGO Forum on Indonesian Development
IP	Internet Protocol
ISD	Initiatives for Sustainable Development toward the Twenty-first Century
ITUT	Centre for the International Transfer of Environmental Techniques, Germany
IUCN	International Union for Conservation of Nature
IWC	International Whaling Commission
JANNI	Japan NGO Network for Indonesia
JARPA	Whale Research Programme under Special Permit in the Antarctic, Japan
JBIC	Japan Bank for International Cooperation
JETRO	Japan External Trade Organisation
JICA	Japan International Cooperation Agency
JODC	Japan Overseas Development Corporation
JP¥	Japanese Yen
KfW	Kreditanstalt für Wiederaufbau, Germany
KITA	Kitakyushu International Training Association
km	Kilometers

LDP	Liberal Democratic Party, Japan
MDG	Millenium Development Goals
MITI	Ministry of International Trade and Industry
MOF	Ministry of Finance
MOFA	Japan Ministry of Foreign Affairs
mt	Million tones
NEDO	New Energy Development Organisation
NGO	Non-Governmental Organization
NIC	Newly Industrialized Country
ODA	Official Development Assistance
OECC	Overseas Environmental Cooperation Centre, Japan
OECD	Organisation for Economic Co-operation and Development
OECF	Overseas Economic Cooperation Fund, Japan
PPP	Public-Private-Partnership
SED	Socialist Unity Party, Germany
SME	Small and Medium-Sized Enterprises
SPD	Sozialdemokratische Partei Deutschlands, Germany
tkm^2	Thousand square kilometers
TZ	Technical Cooperation
UBA	German Environment Agency
UN	United Nations
UNCED	United Nations Conference on Environment and Development
UNCHE	United Nations Conference for Human Environment
UNCTAD	United Nations Conference on Trade and Development
UNEP	United Nations Environment Programme
UNESCO	United Nations Educational, Scientific and Cultural Organization
UNFCCC	United Nations Framework Convention on Climate Change
UNU	United Nations University
USA	United States of America
US$	US-Dollar
VCI	German Association of the Chemical Industry
VDMA	German Association of Local Machinery and Industrial Equipment Manufacturers
WCED	World-Commission for Environment and Development
WDCS	Whale and Dolphin Conservation Society
WID	Women in Development
WSSD	World Summit on Sustainable Development
WWF	World Wilde Fund for Nature
WZB	Social Science Research Centre Berlin, Germany
3R	3R Initiative (reduce, reuse and recycle)

Part A – Approach
1 Introduction

Since the 1990s we are increasingly being confronted with an environmental crisis – one of local, national, transnational and even global dimensions. The destruction of tropical rainforests, the pollution of oceans, lakes and rivers by e.g. oil spills, the exploitation of natural resources, growing mitigation of wastes, the decreasing biodiversity and most prominently climate change are among those issues exemplifying today's problems of a world risk society, not confined within national borders or the frontiers of economic regions. Furthermore, these risks illustrate the extent to which different nations and economies are linked as they are part of one system.

For some analysts this is empirical evidence of states' decline as it demonstrates the breakdown of the distinction between foreign and domestic affairs in an interdependent international system.[2] Arguments for the diminishing of a state's loss of effective control, autonomy and sovereignty are exchanged,[3] focusing predominately on a declining ability of national governments to govern. Increased cooperation efforts among states at the international level, as well as the rise of non-state actors in environmental and development politics are indicators for the erosion of a state's sovereignty. States are no longer perceived to be the only legitimate actors in these fields. In consequence governance should no longer be synonymous with governments, but rather with problem solving ability (Jachtenfuchs 1998).

Accordingly, states must reconfigure the articulation of their own roles, taking those of many other actors into account (Ross 2002). Additionally, more and more analysts grasp globalization and denationalization not only as inevitable destiny, but as a promising opportunity towards problem solving strategies and a call to redefine states' possibilities of intended political guidance and intervention.[4]

The discussion about the problem solving ability of today's states has not deterred from highlighting few national trailblazers through international comparison. One such national trailblazer was Japan in the 1990s – often referred to as being among the economic superpowers as well as among the most progressive states in environmental policy. However the calls of Japanese politicians for

---

2   See for example Morse 1970, Crozier et al. 1975; Sklar 1980, Jänicke 1986, Young 1990 & 1997, Biermann et al. 2002.
3   These are the two central features of the Westphalian "Temple" (Zacher 1992: p. 60f).
4   See for example Luhmann 1988a, Scharpf 1989, and Zürn 1998.

a leadership role in the global environmental arena have fallen silent since the late 1990s due to various domestic structural problems that have moved into the focus of public interest. The government's inability to clarify its own position in the world during the Kyoto Conference for Preventing Global Warming, its lack of leadership and visibility during the United Nations Framework Convention on Climate Change (UNFCCC) fifteenth Conference of Parties (COP15) in Copenhagen in December 2009 (Johnston 2009) has changed the image of Japan as a global environmental flagship. What is more, this lack of position and guidance was pronounced despite the ambitious "Hatoyama Initiative"[5]. Nevertheless, some Japanese companies have started in the 1990s to campaign for the adoption of eco-efficient strategies such as "Zero Emissions" as a new business norm and therefore a pioneering role in the development of environmentally sound products and clean technologies. Special emphasis has been given to the development of alternative fuel and so-called low or even Zero Emissions vehicles.[6] Japan experienced a phenomenal rapid economic growth after the total moral, militaristic and political catastrophe resulting from fascism and militarism. Nevertheless, Japan became an economic superpower with competitive manufacturing industries, allowing it to be strong in exports but relying on effective state machinery. Moreover, Japan is not abundant in natural resources but substantially depends on imports from around the world.

There is consensus among environmental orientated analyses that the industrial society is on the way to severely harm the basis of life; to add salt to the wound, necessary comprehensive preventive measures are not promised at present. As the problems are multiplying, nation-states have proved weak at the sight of these threats. Rapid urban and industrial growth in Newly Industrialized Countries (NICs) has led to changes in the overall pollution intensity. The structural transformation of industry in these countries has brought about marked

---

5   In his address to the 64th Session of the United Nations General Assembly on 24 September 2009, Japan's Prime Minister Yukio Hatoyama officially pledged that Japan would drastically cut greenhouse gas emissions by 25% from 1990 levels by 2020, while proposing to provide vigorous support to developing countries through the transfer of environmental technologies and funding of approximately US$ 11 billion until the end of 2020 to fight pressing problems resulting from global warming.

6   Types of Zero Emissions Cars: Electric, Fuel Cell, Solar, Hydrogen. Types of Low Emission Cars: Hybrid, Liquefied Petroleum Gas, Biofuels, Alcohol, Gasohol, Biodiesel, Methanol.

       The Japanese automotive manufacturer Toyota has sold two million hybrid cars between 1997 and 2008 and reduced greenhouse gas emissions through its vehicles by 7 million tonnes of CO_2 during this time (see http://www.toyota-media.de – 01 August 2010).

changes in the nature and composition of pollution streams and other impacts on the environment.

There is accordingly a growing emphasis in the areas of development and environmental protection to transfer modern technologies which implement the continuous improving of processes, products and services by the conservation of raw materials and energy and by the reduction of toxic substances, waste and emissions within the production cycle. In addition the field of environmental technology transfer reflects the three dimensions of sustainability: (i) a growing market for environmental technology with enormous export potentials, (ii) equal and fair development opportunities relying on the social, intellectual, creative and cooperative skills and aptitudes replacing material resources by an expanded body of knowledge and capacity, and (iii) the reduction or the entire avoidance of negative and harmful effects on systems concerned through human interventions.

The case of environmental technology transfer is a challenge for transnational environmental policies. It involves a wide range of public and private actors, leading to many activities, room to manoeuvre, decision-making processes and partnerships. This case illustrates for a very specific area a development increasingly perceivable in many transnational environmental policy areas. It brings along the questions regarding a nation-state's ability to respond to the challenges of environmental technology transfer and hence redefine its possibilities of intended political guidance and intervention, and a possible transferability of the conclusions to other policy sectors.

Prioritizing environmental technology transfer to NICs as part of its transnational environmental policy, Japan took on these challenges in the 1990s. In consequence the focus of this study is also linked to the often repeated thesis that Japan's Official Development Assistance (ODA) is mainly part of its foreign trade policy (Kevenhörster 1995; May 1989: Nuscheler; Rohde 1995a). Meanwhile, in Japan and Germany the inflationary categorization of technology to be exported as environmentally-sound brings along the question whether all stand firm in a holistic environmental assessment or whether these attempts are to substantially benefit from the growing markets worldwide.

This analysis treats and compares the reciprocal effects between:

- Transnational environmental policies and the effectiveness of national governance under the changed framework conditions through globalization;
- Environmental, technological and societal development;
- Environmental development cooperation and national interests.

While an analysis can take place with the means of sciences without considering political interests, the development of a feasible solution strategy is only possible if a multiplicity of factors such as the social and political inertia as well as the economic interests and social contrasts, different interpretations of the present etc. are taken into account. A political programme for the purpose of problem solving is required therefore not only too much, but even unnecessary for this analysis focusing on the question if and how Japan gained power through its environmental technology transfer to NICs.

Hence, the key questions of this study can be grouped into five broad areas:

1. What were the global and national responsibilities of Japan with regard to the transnational and global environmental problems in the 1990s? How and to what extent did Japan have direct and indirect impact on environment and development in NICs?
2. Who were the national actors and institutions formulating, implementing and evaluating the policies on environmental technology transfer in Japan? How did they do it?
3. What is the net effect of Japan's transnational environmental policies? How and to what extent did the government suffer from a "loss of control" as a result of globalization? Or did it gain control?
4. How did Japan's national government include the practical realization of the paradigm "sustainability" into their policies? Did this paradigm affect the governance ability of national governments?
5. How could Japanese policies be strengthened? What are the future perspectives, and what could Japan learn from Germany and vice-versa?

In order to answer these research questions, this study develops the theoretical framework out of an analysis of discussions, which arose on the continuously increasing undermining of border lines through the growing importance of political, economical and social transnational activities. Albrecht Dehnhard's neo-realistic theory arguing that states are losing power and influence at the international level, but simultaneously gaining power and influence back at other levels forms the theoretical fundament of this study. Thus, it intends to verify that Japan has gained power through its environmental technology transfer to NICs, though this has not resulted in a sustained power due to its policies in other transnational areas. Nevertheless, this analysis concludes that states such as Japan remain key actors in transnational environmental policies.

The theoretical framework of this study requires a holistic, multi-dimensional approach. A combination of an actor's analysis with the policy-cycle model is chosen as the methodological approach. By means of this, this study does not only analyze the relevant actors and approaches in the transfer of environmental technologies to bridge and to gradually remove previous and present weaknesses of transnational environmental policies, but also assess those along the paradigm of sustainable development and a state's problem solving ability.

Wherever possible and appropriate, this study is going to make comparisons with Germany, because both nations share a number of characteristics and have identified the transfer of environmental technology to NICs as one focus of their transnational environment policies. The comparative element with Germany is applied to gain additional knowledge in answering the study's key questions and illustrate certain transferability to other nation-states.

This study consists of four main parts. Part A is a general introduction to the topic, followed by an illustration of the current and global changes as an approach towards the study's focus. A scientific view forms the basis of a paradigm, determining relevant questions to be posed, methods e.g. for the assessment of the policies, partial theories and statements on individual phenomena. Sustainable development is presented as the necessary and commonly agreed-upon paradigm for this research. This is followed by a differentiation between globalization, denationalization and transnationalization as the challenge for the policies. Then, the study approaches a definitory clarification of the key-terms – environment, technology, environmental technology and Newly Industrializing Countries (NICs). The categorization of environmental problems and the resulting challenges for nation-states through those of transnational dimension are in the centre of the next chapter. In a next step the theoretical framework of this study is approached by reflecting the scientific discussions of neo-realists, globalists and others on the changing role of nation-states under transnationalization. This first part is concluded by a presentation of the methodological framework through an actors-analysis, the policy-cycle model (PCM), comparison and the applied methods for undertaking this research.

Part B starts with an analysis of the history of transnational environmental cooperation through treaties, especially looking into the cases of the Whaling Convention and the Kyoto Protocol on the multilateral level, followed by presenting the special quality of bilateral transnational cooperations. With special reference to the required holism of the theoretical approach of this study, the following chapter analyzes general framework conditions and finally the impact on the environment, which are considered of relevance for understanding transnational environmental policies and their respective development.

Part C analyzes the transnational environmental policies through the transfer of environmental technologies to NICs according to the steps of the policy-cycle model: Initiation, Estimation, Selection, Implementation, Evaluation and finally Termination. Within the Initiation special focus is on the development of transnational relations, the phase of ecological ignorance, growing global consciousness, growing public development assistance concluding with the identified political motives for the initiation. The chapter on Estimation is centred on international conferences and resulting institutional changes in the transnational environmental policies, but especially the Official Development Assistance (ODA). The chapter on Selection looks into the Basic Environment and Green Aid Plans, whereas the chapter on Implementation analyzes the institutional framework, the role of consulting companies, local and national agencies, the private industry and Non-Governmental Organizations (NGOs). The chapter on Evaluation is performed by various criteria of effects, guiding to the final step of the policy-cycle model, the Termination.

Part D summarizes the findings, gives answers to the key-questions and draws conclusions out of this study.

2 Global Changes: Approaching the Research Object

2.1 Paradigm – Sustainable Development

The first step in an analysis of transnational environmental policies is to find either an existing theoretical model on which the investigation can be based, or to develop one. A scientific view forms the basis of this paradigm, determining relevant questions to be posed, methods to be used for the assessment of the policies, theories and statements on individual phenomena (Kritz 1990: p. 77f).

Although developmental policies have been the subject of scientific analyses for some 50 years, the discourses in the early 1990s underscored the helplessness at the face of failure of a big or mega-theory (Habermas 1985). It was the German social scientist Ulrich *Menzel* (1991) who called for a development theory, explaining and making clear the economic and social changes of the social and world historic processes. A couple years later it was also Menzel (1998) who came to the conclusion that the enfolding chaos and re-development of most developing countries leaves the development theory without an object and thus meaningless. In Menzel's view the majority of the 180 developing countries are without law and order and thus only caricatures of states. Consequently, the majority of the states supposed to form One World do not really exist in practical terms. This gives reason for Menzel to declare the end of One World and thus the irrelevance of all development theories.

Fundamental reservations against Menzel's thesis appear justified. Even in Africa which is often referred to as a symbol for the chaos-scenario, profound consolidation approaches towards functioning states are obvious. Attempts to track down the secrets of successes in East and Southeast Asia help to identify the functions of cultures, traditions, values, attitudes and behaviours in development and modernization processes. And finally theories are also asked to explain the reasons for states in crisis and failing states. Despite this controversial discussion, there is common agreement that those theories having a universal claim could not resist the different realities in the heterogeneous developing world. They must be considered as failed because they did not succeed in providing convincing explanations and thinking for future developments. This leads to the linkage of explaining strategic or teleological functions of theories. Nevertheless, this is especially important for the discussion of the question which theoretical approach appears most appropriate for this study.

The Modernization Theory and the International Dependence Theory are among those theories which are regarded in the above mentioned debate as failed (Harborth 1992: p. 233ff). Modernization Theory is the fundamental proposition that people in traditional societies should adopt the characteristics of modern societies in order to modernize their social, political and economic institutions and thus to follow the model of developed capitalistic nations. According to International Dependence Theories, the cause of underdevelopment is the dependence on industrialized countries while internal factors of developing countries are considered irrelevant or seen as symptoms or consequences of dependence. They concentrate on explanations of the genesis of underdevelopment and pay little attention to strategies for overcoming the situation.

In addition to the above-mentioned theories, additional were regarded as failed. This includes the Real Existing Socialism and thus a model of development through centralized state planning, in which strategic function is in the foreground. The fourth mega-theory is the Flying Geese Pattern of Development (*gankō keitai hattenron*) formulated by the Japanese economist Akamatsu Kaname.[7] Accordingly, economic development is based on the transfer of outphasing industry and the respective know-how to countries in the slipstream. In applying this model to East Asian development indeed resulted in cascade-like steps based on state-planned private economies. But with the acceptance of the Washington Consensus[8] und the deregulation of their financial markets, Western countries flooded East Asia with surplus capital and thus caused this model to fall (Thiel 2001: p. 13).

Both the Concept of Development as Cultural Phenomenon and the Concept of Sustainable Development go beyond the economic sphere. The essential point of the former is that only analysis of interdependencies outside an economy can shed light on the question why economic factors sometimes have an impact at a certain place and time, but sometimes do not (Weber 2000). Many factors must come together in giving impetus for development. Thus, the Confucianism in Japan and China did not show the same effect for development, because both countries have different value systems. In Japan the existence of a rational bu-

---

7 For more information see e.g. Korhonen 1994, Okita 1956.
8 The Washington Consensus was a set of policies designed in the early 1990s which were believed to be the formula for promoting economic growth in the developing world. These policies called for (1) fiscal discipline, (2) a redirection of public expenditure priorities toward fields offering both high economic returns and the potential to improve income distribution, such as primary health care, primary education and infrastructure, (3) tax reform (to lower marginal rates and broaden the tax base), (4) interest rate liberalization, (5) a competitive exchange rate, (6) trade liberalization, (7) liberalization of inflows of foreign direct investment, (8) privatization, (9) deregulation (to abolish barriers to entry and exit) and (10) secure property rights. This consensus was shaken in the early 2000s, primarily by Argentina which underwent economic problems despite following the consensus.

reaucratism, which Max *Weber* considered as rather important, developed under the pervasive influence of the zen-buddism, whereas China's caste of civil servants was unable to achieve the same for a long time. Communication problems between anthropologists and economists, or more generally between humanists and scientists, concluded that cultural explanations are still facing economical approaches (Klitgaard 1994). Consequently, the potentials of these culturalistic approaches to explain development processes are not yet substantially used.

In contrast to the above-mentioned theoretical approaches, sustainable development is basically a normative proposal. The two German scientists Konrad *Ott* and Ralf *Döring* worked out a theory of sustainability and sustainable development (2004), in which they favour a concept of strong sustainability. A central thesis is the core meaning of the term "sustainability" and thus the long-term and environmentally-sound usage of critical stocks of natural capital. By providing examples of fishery, climate, agriculture and natural protection, they illustrate how strong sustainability can be implemented in practice. But to what extent this approach will find favour among other academics is still open.

Sustainability and the process towards this aim, sustainable development, have become catchphrases used by many different groups to describe greatly varying values and hence a term commonly referred to by financial institutions, governments, companies, civil unions and grassroot organizations all over the world, despite all ideological and political differences (Robèrt 2004: p. 4; Reid 1995: p. 9; Brooks 1992: p. 22). In 2004 the German online dictionary for sustainability listed 36 definitions for sustainability (Aachener Stiftung 2004). This catalogue ranges from commissions of the German Parliament, action-plans of the European Commission and scientific publications of German parties and the Protestant Church. For some, sustainable development has primarily an ecological dimension, so that they interpret it in terms of the environment, resource conservation and the preservation of species. For others, the socio-economic expansion is of equal or higher importance. Yet others refer primarily to economic sustainability when interpreting sustainability. This diversity of definitions highlights the need for precise definitions, so that clear decisions can be made on sustainable development. One of the difficulties in using the concept lies in the fact that it touches on many areas of policy which are generally dealt with separately. It encompasses technical, economic, ecological and cultural aspects, drawing them into one model. This multi-disciplinism makes it difficult to establish an all-encompassing concept.

Nevertheless, all agree that future development cannot follow the model of the past and that ways are needed to achieve economic, social and ecological objectives at the same time. Also considering long-term implications of decisions is felt to be important. Thus, sustainable development has become a magic formula for a global discourse of common interest and the key reference term for future development.[9]

The Limits of Growth (1972) by Donella *Meadows*, Dennis *Meadows*, Jørgen *Randers* and William *Behren* was the first report to the Club of Rome[10] illustrating the danger of industrialized nations to follow the example of the forests just after Nicholas Georgescu-Roegen's *The Entropy Law*[11] *and the Economic Process* (1971), in which he explained that reprocessing requires enormous amounts of energy and must thus be considered as non-ecological. Also, in 1972 the United Nations Conference for Human Environment (UNCHE), the so-called Stockholm Conference, agreed on the establishment of the United Nations Environment Programme (UNEP). Maurice *Strong*, UNEP's first Executive Director, coined the term "eco-development", describing an approach to development aimed at harmonizing social and economic objectives with ecologically-sound management in a spirit of solidarity with future generations. Strong based this on the principle of self-reliance, satisfaction of basic needs, a new symbiosis of man and earth; another kind of quality growth, not zero growth, not negative growth (Eblinghaus & Stickler 1998).

The concept of eco-development is regarded as a precursor to the concept of sustainable development, already reflecting the social, economic and ecologic concerns. The term "sustainable development" was used for the first time in 1980 in the World Conservation Strategy.

With the assignment through the United Nations General Assembly to formulate a worldwide programme of change, the World Commission on Environment and Development (WCED) (1987) further developed the concept of sustainable development towards political implementation, especially in its report *Our Common Future* known as the Brundtland-Report. This report is commonly regarded as the breakthrough, unifying the discourses on environment and development through introducing the term "sustainable development" to a wider

---

9   For more information on the history of ideas on sustainable development see for example Ott & Döring (2004: p. 19-30).
10 The Club of Rome is a group of scientists, economists, businessmen, international high civil servants, Heads of State and former Heads of State who pool their different experiences from a wide range of backgrounds to come to a deeper understanding of the world problematique (http://www.clubofrome.org – 06 December 2004). For more information see e.g. Pauli 1987 & Whitehead 2004.
11 The second law of thermodynamics says that every time energy is transformed from one state to another, there is a loss in the amount of that form of energy, which becomes available to perform work of some kind. The loss in the amount of available energy is known as entropy.

audience. The Brundtland-Report provided the following definition of sustainable development.

> "Sustainable development is development that meets the needs of the present without compromising the ability of future generations to meet their needs." (WCED 1987: p. 48)

Thus, at its heart sustainable development is the idea of ensuring a better quality of life for everyone, now and for the generations to come. Simultaneously, the Brundtland-Report provides key-objectives for action and necessary elements for approaching these goals, which are still adopted in today's discourses and thus considered as benchmarks:

- Effective protection of the environment and endangered species,
- Fighting poverty and satisfaction of basic needs,
- Prudent use of the limited natural resources,
- Social progress which recognizes the needs of everyone,
- Participation and democratization,
- Channelling the population growth in the South.

Moreover, this report and the following discussions have provided a number of key elements forming a concrete action-guide necessary to successfully approach these key objectives.

Economic growth is generally thought necessary to reach prosperity and to fight e.g. poverty and unemployment. But economic growth with present production technologies leads to the waste of limited resources and environmental pollution. Technological and social innovations and a re-grouping within and among economic sectors could help to decouple Gross Domestic Product (GDP) from the consumption of natural resources and environmental pollution. Consequently, an increased resource-efficiency is considered as mandatory to minimize the negative environmental impacts of the economic system (Lehner & Schmidt-Bleek 1999: p. 157ff; v. Weizsäcker et al. 1996: p. 302ff; Hawken et al. 2000: p. 92). The costs through consumption of natural resources and the environmental pollution in production, consumption and disposal should be adequately reflected in the final price (Bund & Misereor 1996).

Good Governance is regarded as another key-element for approaching sustainability. This implies governments legitimated through elections, transparency in decision-making of the administration, rule of law and freedom of press and speech, being a pre-requisite for a fair spreading of goods within a democratically framed market-economy. These are examples for action guiding elements in approaching sustainable development. Two more are in the focus of this work and will be analyzed in-depth in the following: (1) transnational environmental policies and (2) the transfer of environmental technologies.

With its report the WCED formed the basis for the United Nations Conference on Environment and Development (UNCED). This so-called Rio Conference anchored the concept of sustainable development in the Agenda 21 and the Declaration on Environment and Development in 1992, which was subsequently signed by 172 states. Hence, the international community agreed on sustainability as the paradigm after two decades of intensive preparations. Still, during the World Summit on Sustainable Development (WSSD) in Johannesburg in 2002, also called Rio+10, it was hardly a point of dispute that progress in implementing the action-plan towards sustainability had been extremely disappointing, with poverty deepening and environmental degradation becoming worse (UN 2004).

In Germany the concept of sustainable development developed out of environmental considerations in the economic usage of forests and thus became a synonym for environmentally sustainable economic development. Whereas in Japan it was first translated into *jizokuteki kaihatsu* or *jizoku kano na kaihatsu* and then into *jizokuteki hatten*, which simply means continuous development or economic development, omitting the part on environment completely (Fuwa 1999). Consequently, the original meanings of sustainable development differ in Japan and Germany and it is a task for this study to find out whether there are certain indicators in the respective transnational environmental policies of both nations as evidence for different understandings.

Nachhaltige Entwicklung Dauerhaft umweltgerechte Entwicklung Ökologisch-dauerhafte Entwicklung Zukunftsverträgliche Entwicklung Zukunftsfähige Entwicklung	持続的開発 持続可能ナ開発

Figure 1: Japanese and German expressions of sustainable development
(Source: Own Illustration)

Over a long period of time, the public debate on sustainable development has been with different foci, partly confrontational and often fragmented. Problems are often addressed one by one as they occur and attract public interest. Chernobyl, climate change, ozone depletion, the destruction of tropical rainforests, the pollution of oceans, lakes and rivers, BSE, Foot and Mouth Disease, SARS, the avian bird and swine flu... until something else attracts more interest. Unfortunately, relationships between issues are rarely considered which is also why the United Nations General Assembly reaffirmed after the WSSD in December 2002 the need to ensure a balance between economic development, social development and environmental protection as interdependent and mutually reinforcing pillars of sustainable development. The UN General Assembly also restated that poverty eradication, changing unsustainable patterns of production and consumption, and protecting and managing the natural resource base of economic and social development are overarching objectives of, and essential requirements for, sustainable development (UN General Assembly 2004).

Since the Rio Conference the majority of countries committed themselves in setting up national advisory councils to promote dialogue between governmental representatives, business people, environmentalists and others actors on sustainable development policies.[12] In 2002 it was reported that almost 90 national strategies aiming towards sustainability had been passed, although substantially differing in its contents and outreach (Geiß 2002: p. 117).

The Government of Japan presented its National Action Plan for Agenda 21 already in 1994, whereas an Enquete-Commission of the German Parliament delivered its final report entitled Concept Sustainability: From Paradigm to Implementation in 1998. However, Germany's government only agreed in April 2002 on the national strategy towards sustainable development (Government of Japan 1994; Government of Germany 2002). Many cities and towns have since created their own Local Agenda 21. In 2001 it was reported of decisions of a Local Agenda 21 in about 2,300 German cities and towns (Ruschkowski 2002: p. 21). Ten years after the Rio Conference, the Johannesburg Summit resulted in the Plan for Implementation and in the launch of more than 300 voluntary partnerships (UN 2004). Each of these is attempting to bring additional resources to support efforts to implement the concept of sustainable development. Thus, these new partnerships, tied to government commitments, provide a built-in mechanism to ensure further implementation. Nevertheless, most NGOs indicated that the outcomes of the WSSD were too vague, leading to a sense of doubt of a new era about to dawn after Rio.

Despite these obvious problems, the concept of sustainable development has become the paradigm for policymaking in most countries, at least by lip-saying and action-plans. In addition there is general agreement that this model provides a possible way out of the dilemma mega-development theories have left behind. Besides, sustainable development is normative, teleological and containing both a strategic and explaining function. Consequently, the concept serves as a para-

---

12 See for example International Institute for Sustainable Development et al. (2004), which analyzes with its partners national strategies and policy initiatives for sustainable development used by a number of developed and developing countries.

digm for the purpose of this investigation on Japan's transnational environmental policies. It will help in the assessment of the respective policies under analysis against the theoretical basis of this study, but also against the claim of a nation and what was agreed on worldwide.

2.2 Transnationalization – The Challenge

In recent years globalization has become another catchword for the present course taken. This new area of debate is regarded not as a threat, but as an opportunity for a better and fairer livelihood if successfully combined with sustainable development.

From the persistent discussions about the role of national policies during an era of transformation the questions arise not only where the discrepancies between potentials and perspectives of states are and which approaches are promising to manage the problems effectively in the future, but also what these transformations are and how they are influencing the national policy formulation process.

"Globalization" became the key-term for those changes and thus also a key-term of scientific discussion within the last years. Anthony *Giddens* defines "globalization" as the:

> "...intensification of worldwide social relations, by which distant places are interconnected in such a way that events at one place are influenced by processes, which take place many kilometres away and vice-versa."[13]

This definition refers to the increasing interconnectedness of economies and markets through

- A geographical enlargement and growing interactions' density of international trade including the rapid increase of exports from newly-industrializing countries;
- The global networking of financial markets, resulting in rather flexible capital flows escaping from national tax authorities including the growing direct investments in other countries through multinational corporations;
- New transportation and communication media abolishing spatial separation.[14]

The German philosopher Karl *Jaspers* described these characteristics of globalization already four years after the Second World War in his book on the origin

---

13   Translated by the author from German into English. See Giddens 1997: p. 85.
14   See also Habermas 1998: p. 70; Robertson 1992: p. 8.

and target of history. He portrayed the so-called technical epoch as absolute universal (Jaspers 1949: p. 178f). More recently, in sociology, it has been argued that

> "(...) each major aspect of social reality (the structure, culture and personality of traditional terminology) is simultaneously under-going globalisation, as witnessed by the emergence of a world economy, a cosmopolitan culture and international social movements" (Archer 1990: p. 1).

The principal elements of globalization were succinctly enumerated already more than three-quarters of a century ago from today. For instance, the Communist Manifesto stated:

> "Constant revolutionizing of production (...): The need of constantly expanded market for its products chases the bourgeoisie over the whole surface of the globe. It must nestle everywhere, settle everywhere, and establish connections everywhere."(Roy 1995: p. 2,005)

But still for some, globalization signifies more than this. The term is often seen as synonymous with modernization, as new techniques are manufactured in global supply chains and flow across the globe with increasing economic opportunity and efficiency, as global exchange replaces fragmental local and national markets. Thus, for a long time globalization was regarded as the realization of the eternal prosperity. However, literature assumes a loose and overstretched notion of the term "globalization" with both quantitatively and qualitatively different phenomena in environmental, economical and social relations (Clark 1998: p. 479).

Some argue that the term "globalization" would be misleading and therefore plead for the introduction of "denationalization" instead. The German social scientist Michael *Zürn* argues that in addition to the trade-flows, additional cross-border transactions must be taken into consideration for an appropriate measurement of societal denationalization:

- Violence = Exchange and production of weapons and threats;
- Communication & Culture = Exchange and production of songs and cultural products;
- Mobility = Travel and migration;
- Economy = Exchange and production of goods, services and capital;
- Environment = Exchange and production of harmful substances and risks.
  (Zürn 1998: p. 71)

In view of the dynamic and non-static processes of current transformations and the multi-layered complexity, a renaming into societal denationalization appears only appropriate with Zürn's approach to operationalize the definition taking the immense social and environmental dimensions into consideration. However, a certain quality must be reached, the same with globalization. The simple increase in cross-border transactions is not necessarily an indicator that results in worldwide impacts and thus on a global scale, but certainly on a transnational level.

Table 1: Geographic distribution of Internet Protocol[15] Locations, (2007)

Geographic area	Number of addresses	Percentage
Africa	40,241,664	1.519%
Antarctica	15,620	0.001%
Asia	371,297,015	14.015%
Caribbean	1,681,866	0.063%
Central America	2,557,340	0.097%
Europe	569,838,903	21.510%
Middle East	12,011,131	0.453%
North America	1,481,754,661	55.932%
Oceania	76,417,711	2.885%
South America	93,409,304	3.525%

(Source: IPligence 2009)

Furthermore, discussions about a globalized world refer to the worldwide access to the Internet and exchange of emails. However, the worldwide distribution of Internet assets illustrated in Table 1 demonstrates that in 2007 North America and Europe concentrated the two largest groups with a total share of 22.5% for Europe and more than twice this amount for North America – 55.9%. It can therefore be said that that these two regions represented more than three-fourths of the global Internet structure.

Figure 1 shows that regions with the highest population e.g. certain regions of Africa, India and China, are behind in Internet adoption. Africa and Asia, having only an estimated 4% and 10% of Internet penetration, respectively compared to the 70% penetration in the United States.

Hence, in the early 21st Century the Internet has failed to reach most parts of Africa, South America, the Middle East, South Asia and major parts of East Asia. Because of this lack of an overall evenly, exhaustive distribution, it is difficult to characterize our world as a "globalized world", a status, instead of the process towards this – "transnationalization" or "globalization", recognizing the speedy change of this situation.

---

15 Internet Protocol (IP) is the principal formal description of digital message formats and the rules for exchanging those. It is used for relaying datagrams (packets) across an internetwork.

2 Global Changes: Approaching the Research Object 17

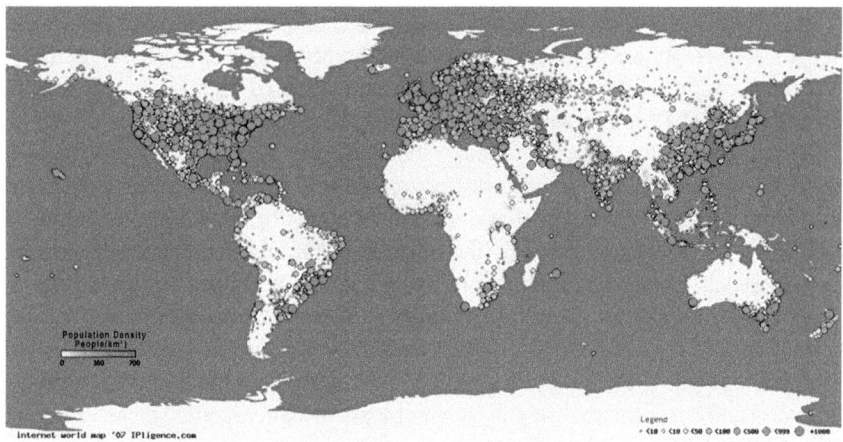

Figure 2: Population Density & Internet Distribution (2009)
(Source: IPligence 2009)

There are many examples that the world is transnationalizing and occasionally globalizing, in the sense that activities take place increasingly across borderlines not only in certain regions or between certain countries, but all over the world and on a supranational or transnational scale. A snapshot of the worldwide distribution of Internet connections cannot be used as an argument against the use of the term "globalization", because it clearly illustrates the cross-border exchange and production of communication and thus supports Zürn's arguments towards the usage of the term "societal denationalization". Yet only the increase of worldwide communication does not necessarily touch the role of nation-states, so that this is evidence for a societal denationalization. Therefore certain quality thresholds must be exceeded in the five above given areas, which would verify the relative increase of intensity and range of cross-border exchanges and production processes (Ibid: p. 76). Denationalization is a variable for a process similar to transnationalization and globalization, which can have different levels depending on the focus area and the country under investigation.

In recent years many examples confirmed that in international relations some players interfere from the outside into what is usually considered the internal affairs of a nation. A few evident crises include the 1997 financial market crisis in the Far East, Russia and South America but most recently the world financial crisis demonstrate the frightening dimensions of decreasing significance of space and time through those changes. Moreover, the diminution of the power of nation-states and corporations to manage and control such events became obvious when experts at The World Bank and the International Monetary Fund (IMF) arrived in e.g. Thailand, Indonesia, South Korea and Russia in 1997. Within a couple of days they prescribed a new national budget, the cut of subventions and the (de)regulation of prices and the pass of flanking acts.[16] Under different circumstances governments, including big economic powers such as the USA and Germany, received instructions how to preserve its people and how to secure prosperity and economic growth through e.g. an immense cutting of subsidies for German farmers by the European Commission. But this cannot be considered as "interstate" interactions along with conventional diplomatic activities. The political-legitimated protagonists of nation-states had to carry out the stipulations of representatives of NGOs and thus turned to become a recipient of orders (Ashcroft 1998; Beck 1998: p. 37). Although some global interactions are initiated and sustained entirely by governments of nation-states, these examples clearly demonstrate that some interstate interactions may involve governments, but it may not involve only governments, as non-governmental actors can also play a significant role.[17]

Again, in order to justify the usage of denationalization, these cross-border transactions must have increased substantially and sustainably within a given time frame or must have reached a high level (Zürn 1998: p.76). If this is the case in the above given examples, it is not in the interest of this work. This study attempts to inspect if and how non-governmental, private and/or international actors are influencing how it is with Japan's transnational environmental policies through environmental technology transfer.

Consequently, interstate interactions, however, involving non-governmental actors – individuals or organizations – are considered from a political-scientific point of view as "transnational" (Menderhausen 1969; Risse-Kappen 1995: p. 3). And based on the growing importance of non-governmental actors in the field of international environmental policies and development policies,[18] and especially also in the case of environmental technology transfer through private-

---

16 See reporting in all kinds of media during the visits of IMF representatives to those nations (e.g. Japan Times 29 April 1999, "NHK 9:00 o'clock news" 12 May 1999).
17 In response to numerous international scandals regarding hazardous waste trafficking that began to occur in the late 1980s, international NGOs such as Greenpeace and the Basel Action Network (BAN) have a substantial part in the Basel Convention on the Control of Transboundary Movements of Hazardous Wastes and their Disposal, which was adopted in Basel, Switzerland on 22 March 1989. Even during the respective Conference of Parties (COP) one hesitates to push through decisions against the strong opposition of BAN, the self-defined watchdog of the Basel Convention, which usually succeeds to back up its points with some parties.
18 See e.g. Weiß (2000).

public-partnerships (PPP), it also appears justified to speak from an ongoing "transnationalization".

> "Transnationalisation is characterised through the growing role individuals and organisations play in world politics vis-à-vis foreign government or foreign societies and thus bypass their own governments." (Singer 1969: p. 24)

Globalization, denationalization and transnationalization processes perforate border lines drawn by nation-states, which finally lead to a polycentric distribution of power addressed by George *Rosenau* in 1990. This means a turning away from the sovereignty of Hobbes's *Leviathan* and Herz's *Hard Shell*, which understand the statesperson as the only source of governance represented through a government (Hobbes 1996: p. 144; Herz 1976: p. 100). It also explodes Max Weber's understanding of international policy, which implies two things: First that the space in between two nation-states is the sphere of action of politics; and second that the main actors are nation-states. But this kind of politics has never existed in its pureness (Kaiser 1969: p. 81). Additionally, it also reflects one key error in the current globalization debates, namely the identification of the modern state with the nation-state.

Consequently, the role of governments and nation-states as described by e.g. Bodin, Rousseau, Hegel and Morgenthau should be changed. But does transnationalization necessarily lead to a constant loss of control and the states' ability to govern, so that a level is reached which give evidence to discuss denationalization?

The current phase of transnationalization is dominated by economic and financial processes driven by the mobility of capital and the free flow of information (Goldblatt 1997: p. 269). We are witness to the nearly complete disappearance of centrally planned economies and powerful trends towards the use of market forces and market-based policies are evident throughout the world (Robinson & Tinker 1997: p. 71). Although the phenomena of economic globalization can be evaluated in various ways, it remains that economic players dominate this trend, not states. Expanding international markets cause governments and state institutions to face narrow choices in dealing with liberalization. Trade liberalization is driving global economic integration. The turnover at stock exchanges characterizing global financial flows dwarfs flow of traded goods and services. However, the other side of the medal shows that levels of absolute poverty and income disparity are not decreasing around the world (Chen & Ravallion 2008; UNDP 2004; Ibid. 1998). Consequently, current economic conditions must be considered as unsustainable for a large and growing proportion of the world's population, including the industrializing countries. Thus, social structures are put under unprecedented pressure. Socially, the decreasing ability to address basic social issues such as unemployment, pensions for retirees, health care, homelessness, crime, drugs and poverty causes an alienation from and even distrust of the present system of governance. Moreover, the impact of economic globalization is also compounded by further vulnerabilities within national jurisdiction, cultural or environmental terms, for e.g. where the issues of an internationalizing media and technology transfer pose specific culture, policy and public management issues. In the eyes of some analysts, this vulnerability re-emphasizes the possible, newer importance of boundaries and sovereignty within the new international orders (Kouzmin & Hayne 1999: p. 1).

Considering the transnationalization of environmental policies as a test-case for the effectiveness of national initiatives, the analysis of Japan's approaches, also through comparison with Germany, will not only work out differences and similarities, but also angles for possible cooperation and further problem solving attempts. Moreover, the number of comparative analyses responding to various declarations of some of Japan's and Germany's leading politicians – figuratively speaking – to become the head goose of flying geese in global environmental policies is rather limited.

Though one can identify an increasing number of research projects on the modes of governance, analyzing capacities for environmental policymaking and exploring means to support an integration of environmental aspects into other policy dimensions,[19] comprehensive studies analyzing today's nation-states' role in transnational environmental policymaking have not yet been published.

The recent work of the sociologist Manuel *Castells* (2010, 2004, 2000) brings together the many facets of the multitude of changes occurring since the development of the transnational information economy or the Information Age in one panoramic expanse. Castells also deals with the decline of sovereign states and the emergence of new bases of power, in which nation-states are just one of these (Castells 2004: p. 356). Surprisingly, Castells does not refer to business and industry as part of the actual operating unit of political management in a globalized world, which in his view is only formed by nation-states, international institutions, associations of nation-states, regional and local governments and NGOs (Ibid. 2004: p. 364). But as none of the above-mentioned actors own environmental techniques, industry and business players do play an important role in policies supporting their transfer. They cannot be forgotten in this study.

---

19 See for example the projects of the Environmental Policy Research Centre at the Free University of Berlin, Germany (http://www.polsoz.fu-berlin.de/polwiss/forschung/ systeme/ffu/projekte/ abgeschlossene/index.html – 13 May 2010).

György and Ute *Széll* (2009), Klaus *Vollmer* (2006), and György *Széll* and Ken'ichi *Tominaga* (2004) have put certain foci in their books on environmental policies and ecological issues. Whereas Vollmer's focus is on Japan and East Asia through a transnational perspective, György and Ute *Széll* and Ken'ichi *Tominaga* direct their focus on Japan and Germany, partly in direct comparison or the development of sustainable societies. All are illustrating the multi-facets of policies in response to present environmental challenges, though none of the contributors take up the challenge to analyze the interdisciplinary field of environmental technology transfer.

Hidefumi *Imura* and Miranda *Schreurs* (2005a) focused their book on Japan's Environmental Policy, in which Hidefumi *Imura* together with Ryota *Shinohara* and Koji *Himi* examined Japan's environmental industries and technologies, however only touching briefly on questions associated with the transfer of such (Imura 2005b: 276ff).

The multitude of aspects associated with the transfer of environmental technology ranges from environmental considerations in order to ensure an export of environmentally sound techniques; the existence of appropriate markets; the development of human capacities and capabilities to use and maintain these technologies; the role of nation-states in these transnational policy fields, in which their policy measures primarily support industries and business as owner of the equipment and know-how etc. These various elements make environmental technology transfer a new and challenging field for a policy analysis to which this study aims to contribute.

3 Definitions – The Key Terms

The transnationalization and globalization of environmental issues is fundamentally different from economic and social relations. The global trend in economies and societies is still nascent, with local or regional factors still dominating most economic and social sectors (Dolzer 1998: p. 158). Contrary to economic and social globalization, no one really welcomes the growing globalization of environmental problems as they are viewed as posing a threat to the living conditions of present and future generations. Ecologically, today's industry, agriculture and use of renewable and non-renewable natural resources are undermining the environmental resource base. But the perception and responses to those problems varies considerably between different cultures and nations. Thus, it appears necessary for this study not only to find operational definitions for some key-terms, but also to deduce conclusions for the formulation and implementation of transnational environmental policies.

3.1 Environment

At the beginning of the 20th Century, the biologist Jakob Johann *von Uexküll* introduced the German term *Umwelt* (Japanese: *kankyō*; English: *environment*) to describe the animated and unanimated world perceived by living beings (von Uexküll 1964). Around 1970 "environment" and its equivalents in other languages entered the colloquial languages and since then it is often combined with additional terms, e.g. environmental protection, environmental awareness, environmental care, environmental technology etc., and further attributes e.g. social, economical, cultural, urban, national, global, psychological and last but not least industrial environment. This leads to a certain haziness of the meaning and substance of the term "environment". Simultaneously, the concept "environment" has changed over time from a purely scientific term to encompassing an ecological realm with concern to the protection of animals and natural conservation, to the increasing pollution through industrial processes, such as cleansing approaches to minimize harmful effects. Since the late 1980s/ early 1990s this discussion has entailed sustainable development.

After a period of rapid economic growth after the Second World War, the broad masses became aware of the extent of deterioration and destruction of the environment. The German and Japanese expressions for "environment" primarily referred to pollution in the public and in the media (Japanese: *kōgai*[20], Ger-

---

20 The characters which form the word, ko meaning public, and gai meaning harm suggest that pollution was that which caused harm to the public.

man: *Umweltverschmutzung*). The introduction of environmental protection in response to human activities gained a foothold in Europe and Japan prior to the first environmental conference of the United Nations in Stockholm, Sweden in 1972. The preparations leading up to the conference and its outcomes placed emphasis on the comprehensive nature of environmental problems which the Japanese word *kōgai* did not incorporate. Although the Japanese word *kankyō* was only known to natural scientists, especially biologists and ecologists, the Japan Environmental Agency, established in 1971, was named *kankyōchō* and not *kōgai boshichō* (Pollution Prevention Agency). This choice in nomenclature can be understood as a signal for a more comprehensive understanding of the environment. When the Federal German Government announced its first environmental programme in the autumn of 1971, the pure cleansing approach became obvious (Beckmann 1984: p. 56). This governmental appeal was not taken seriously by the entrepreneurs, so that in 1976 the President of the Federal Environment Agency of Germany, Heinrich *von Lersner*, called it "a pure Sisiphuswork"[21].

In contrast to this, public movements forced the Japanese government and industry to the formulation and implementation of a technocratic environmental policy, focusing on end-of-pipe technologies, and led to a striking improvement in some areas of emissions (Weidner 1996: p. 160f; Tsuru & Weidner 1989; OECD 1994). However, the fast successes of Japan's environmental policy regarding industrial pollution resulted in a satisfaction effect causing the environmental discussion to lose much of its socio-political explosiveness. Fortunately, the environmental discussion was revitalized among politicians, bureaucrats and entrepreneurs in the late 1980s when Japan was often criticized for its negative environmental impacts outside its archipelago (Kühr 1997: p. 202). It was only around this time that the change in vocabulary from *kōgai* to *kankyō* became widely accepted.

In the mid-1980s Germany was confronted with one environmental scandal after another; dioxin, smog and acid rain are some prominent examples. Citizens' initiatives developed as an instrument for forming and expressing interests fighting the above-mentioned and to complain about the defects in various spheres of life, including the environment (Sarkar 1993: p. 30). Their demand for "quality of life" also reintroduced the environmental discussion to the broader public and supported the introduction of a more holistic perspective under the term "environment".

---

21 He said: "By the time we have brought one pollutant under control, another one has become a problem." (Der Spiegel, No. 40; 1976: p. 62).

Both in Japan and Germany, the scientific discipline "ecology" is often defined as environmental science and the majority of environmental scientists have for a long time understood "environment" in its core as a natural-scientific concept. This also resulted from the fact that the term "environment" generally describes ecosystems, i.e. functional ecological communities and their non-living environment, with which they are interrelated and constantly interact with each other.

With regard to mankind environment describes the factors determining and influencing its life. These are the forces, which influence the mental, physical, technical, economic and social existence. An anthropogenic environment can be distinguished from a sociological, spatial and biological understanding. The first one stands for the social milieu of a single human-being, a partnership, a group and even a society. The spatial surrounding of human-beings up to the landscape types is covered by the spatial definition of environment. The biological term contains the status of the environment for animals, human-beings, plants and microorganisms as well as the conditions for its living together and interrelations (Wicke 1982: p. 5; Wittkämper 1992: p. 8).

Although the above definitions may be clear and convincing, the meaning of "environment" is still relatively unclear in politics and to the public, not only in Japan and Germany but also elsewhere. Even experts are of different opinion. One reason is the fact that environment can be structurally or functionally understood (Lovelock 1993: p. 2; Rat von Sachverständigen für Umweltfragen 1987).[22] Unconsciously, environment is often applied to mankind, but the discussion regarding sustainable development shows that it became very narrow and thus isolated from vital issues of everyday life such as workplace safety, healthy communities and food security that are often viewed separately as industrial, community and agricultural concerns (Gottlieb 2001). A holistic concept of sustainable development shows that such a fragmented approach prevents awareness that such issues are also environmental issues.

Although the biological environmental demands are relatively similar around the world, traditions and cultural developments and certain interests led to different subjective environmental expectations, which are obvious in the different concepts e.g. of politicians, entrepreneurs, representatives from developing countries etc. And even at the sight of differences between e.g. the environments of Mid-Europeans and Yamomami Indians and even Germans and Japanese, stumbling blocks toward characterization of "environment" become apparent. As it will be shown in the following description, the necessity to find

---

22 For a more detailed description of the complexity of the term environment see e.g. Haber 1992.

arrangements with the given natural conditions is certainly also of relevance for the development of the environmental understanding of a people or nation. This makes strategies regarding transnational environmental problems difficult.

3.2 Technology

This short introduction into the complexity of the term "environment" makes clear that one has to find an operational definition for the pair of terms "environmental technology", which is in the centre of analysis within this study.

The term technology (Japanese: *kagaku gijutsu* or *tekunorojī*; German: *Technologie*) is a combination of the Greek *technē* meaning "art, craft" with *logos*, "word, speech" meant in Greece a discourse on the arts, both fine arts and applied arts. When it first appeared in English in the 17th century, it was used to denote a discussion of the applied arts only, and gradually these "arts" themselves came to be the object of the designation. By the early 20th century, the term embraced a growing range of means, processes and ideas in addition to tools and machines developed under industrialization. Thus, by mid-century technology was defined by such words as

"the means or activity by which man seeks to change or manipulate his environment."[23]

In the relevant literature on environmental problems, the terms "technique" and "technology" are often used synonymously. However, techniques are in contrast to technologies methods of creating new tools and products of tools and the capacity for constructing such artefacts (Ropohl 1991: p.10). Thus, it is also a determining characteristic of man-like species. Other species also make artefacts, e.g. bees build hives to deposit their honey, birds make nests and beavers build dams. But these attributes are the result of patterns of instinctive behaviour and cannot be varied to suit rapidly changing circumstances. Human-beings, in contrast with other species, do not possess highly developed instinctive reactions but do have the capacity to think systematically and creatively about techniques. They can innovate and consciously modify their environment in a way no other species has achieved yet.[24]

In contrast to the definition of "techniques", "technologies" imply the know-how required to develop and apply techniques and technical procedures (Ham-

---

23 See http://www.britannica.com/eb/article?eu=115396&tocid=10381#10381.toc – 06 September 2010.
24 An ape may on occasion use a stick to beat bananas from a tree: a man can fashion the stick into a cutting tool and remove a whole bunch of bananas. Somewhere in the transition between the two, the hominid, or the first manlike species, emerges.

mann & Mittag 1986: p. 225). Thus, it exists embodied in machinery and equipment and unembodied in blueprints, technical instructions, manuals etc. (Hillebrand et. al. 1994a: p. 2). Additionally, at least two more factors have to be taken into consideration to deploy technologies:

(i) The qualification of the person who operates technologies,
(ii) The organization, i.e. the integration of a given technology into social contexts and operation.[25]

Consequently the term technology reflects four different dimensions as summarized by Hillebrand (1994b: p. 4):

1. The specific configuration of techniques and thus machinery and equipment designed for the production processes or for the provision of services, which can be summarized under the term "technical hardware",
2. The scientific and technical knowledge, formal qualifications and experienced based knowledge, which can be entitled "know-how",
3. The management methods used to link technical hardware and know-how, known as "organization", and
4. The physical good or service emerging from the production process – the "product".

In using its rational faculties to devise techniques and modify his environment, mankind has tackled problems other than those of survival and the production of wealth with which the term technology is usually associated today. The technique of language, for example, involves the manipulation of sounds and symbols in a meaningful way; the techniques of artistic and ritual creativity represent other aspects of the technological incentive. Nevertheless, within the complex relationship between the ecosphere and the anthroposphere, technology presents a link between natural and human exploits.

3.3 Environmental Technology

3.3.1 General Understanding

Humanity is confronted with the problem of limited natural resources and negative impacts through its activities on the environment. Consequently, there is no

---

[25] Technology often embodies organizational factors, i.e. the involvement of organizational knowledge on the possible arrangements of the connection different manufacturing steps e.g. in the chemical industry to close loops.

doubt that it must find new ways to minimize the environmental burden and to make future development sustainable.

Accordingly, there is a growing emphasis in the areas of development and environmental protection on modern technologies to make efficient use of available resources and to reduce environmental impacts. Industry and applied sciences are strengthening their efforts towards cost-minimization through novel technologies, especially when substantial saving potentials through total-quality-management and just-in-time systems are no longer obvious. Furthermore, environmental pollution of local, regional and global dimensions through e.g. industrial processes with partly disastrous direct impact on human life, demonstrated the necessity not only for improvements, but also to find new production and development paths especially taking the paradigm of sustainability into consideration.

This influenced the development of novel machineries and techniques, not only concentrating on production of goods and services in a most efficient and effective way, but also taking possible negative and possible positive impacts on the environment into account. These new requirements are met by technologies, for those designation the combination of the terms environment and technology appeared appropriate, so that they finally became entitled "environmental technology (ET)" (Japanese: *kankyō gijutsu*; German: *Umwelttechnologie*) or sometimes synonymously "environmentally sound technologies (EST)".

Notwithstanding the ubiquitous usage of environmental technology in the literature, an operational definition is not plain (Halstrick-Schwenk et. al. 1994: p. 20; Kuehr 2007). Rath and Herbert-Copley (1993) argue that in the case of environmental technology, environmental impact depends on the way it is used. Yet the same applies for every kind of technology; therefore this approach does not appear very useful. In Förster's (1995) understanding

"Environmental Technology combines technology with natural resources".

This definition is not only too general to be useful in the context of this work, but also misleading. All kind of technologies are directly or indirectly networked with natural resources through their utilization in production and a fundamental aspect of technology is that it requires natural resources.

3.3.2 International Attempts of Definition

From an idealistic point of view, it is desirable that every kind of technology should be environmentally sound and sustainable. This, however, does not correspond with the present reality or the short- and mid-term expectations. The classification "environmental technologies" requires a common agreement of an operational definition. If every actor applies his own standard and understanding, the formulation and implementation of policies along the paradigm "sustainability" or in support of the targets of the Millennium Development Goals[26] must be questioned.

The United Nations Conference on Environment and Development (UNCED) provides a more specific definition of environmentally sound technology:

> "34.1 Environmentally sound technologies protect the environment, are less polluting, use all resources in a more sustainable manner, recycle more of their wastes and products, and handle residual wastes in a more acceptable manner than the technologies for which they were substitutes.
>
> 34.2 Environmentally sound technologies in the context of pollution are "process and product technologies" that generate low or no waste, for the prevention of pollution. They also cover "end of the pipe" technologies for treatment of pollution after it has been generated.
>
> 34.3 Environmentally sound technologies are not just individual technologies, but total systems which include know-how, procedures, goods and services, and equipment as well as organizational and managerial procedures. This implies that when discussing transfer of technologies, the human resource development and local capacity-building aspects of technology choices, including gender-relevant aspects, should also be addressed. Environmentally sound technologies should be compatible with nationally determined socio-economic, cultural and environmental priorities."
> (UNCED 1992: p. 2)

This definition reflects the understanding of the United Nations Environment Programme (UNEP) and its International Environmental Technology Centre (UNEP/IETC) in Japan.[27] According to UNEP "environmental technology" or "environmentally sound technologies" encompass technologies that have the potential for significantly improving the environmental performance relative to other technologies.

In the eyes of the OECD, this definition mirrors the concept of "cleaner technologies". Consequently, technologies which are cleaner than conventional can be categorized as ET (OECD 1995). Through the introduction of these technologies, the core production technology is modified so that emissions and the

---

26 Environmental technology related targets of the Millennium Development Goals (MDG) can be found under Goal 8 "Ensure Environmental Sustainability", namely Target 1 "Integrate the principles of sustainable development into country policies and programmes and reverse the loss of environmental resources" and Target 3 "Halve, by 2015, the proportion of the population without sustainable access to safe drinking water and basic sanitation" (see http://www.un.org/millenniumgoals/environ.shtml – 03 April 2010).

27 http://www.unep.or.jp/ietc/index.asp – 03 April 2010.

consumption of energy and natural resources are reduced. Thus, they lead to higher resource efficiency with financial and economical benefits. By analogy with Agenda 21, the OECD includes goods, services, systems, technical know-how and organizing and management capabilities in its definition. The OECD does not include end-of-pipe or cleansing technologies to ET, because of their high production costs without necessarily increasing the overall production. Additionally, such technologies may only shift environmental problems from one sphere to another one.

Due to the lack of standards of comparison, many elementary questions remain unanswered. The statement that ET is less polluting and recycles more than other technologies leaves much space for interpretation. The same counts for the finding that utilization of resources is more sustainable. But more sustainable than what? What are the indicators?

The prognosis of the growing market of ET led to the integration of ET into the portfolio of countless companies and the establishment of those around the world.[28] The China Greentech Initiative (2009) estimates the annual need of ET to be between US$ 500-1,000 billion, though the actual demand is substantially smaller. The Indian government plans investments of EUR 9 billion until 2012 just for water supply and sewage treatment in 63 cities (German Trade & Invest 2010). Moreover, the majority of them do not define ET or provide such an insufficient description of ET that allows categorizing most technologies in the subgroup ET.[29]

Unfortunately the attempts of UNCED and the OECD do not appear appropriate for either a definition of ET or as basis for this study. Instead it seems

---

28 According to estimations made by the OECD the market for EST was estimated to have topped US$ 300 billion by the year 2000. Of this amount US$ 245 billion was estimated to fall to members of the OECD, with the remaining US$ 55 billion to be made up by non-members (OECD 1992: p. 15). The International Finance Corporation expected an even greater rise, to US $600 billion by the year 2000 (Moore 1994a: p. 236). A study by the Helmut Kaiser Consultancy expected US$ 350 billion to be reached worldwide in 2000 (HKU 1997: p. 4). Today, the world market for environmental technology is estimated to EUR 250-1,000 billion per year (Nordisk Innovations Center 2010; Mallett 2010; German Environment Agency 2007). These massive divergences of ET market estimates result from different categorization of ET and/or the total abandonment of a definition of ET. According to the European Environment Agency (2010) the European Union's eco-industries employ more than two million people, account for about one-third of the global market and are growing by around 5% annually.

29 See http://www.utt-gmbh.de; http://www.gruenenwald-ag.ch; http://www.sternad.com; http://www.jessberger.de; http://www.ecos-consult.com; http://www.envicom.com; http://www.eco-web.com; http://www.etcentre.org – all 15 July 2010.

worthwhile to focus on two institutions in Japan and Germany specialized on the transfer of ET to Newly Industrializing Countries.

3.3.3 Environmental Technology Transfer Institutions

Within Germany's efforts to protect and preserve the environment for coming generations, the Centre for the International Transfer of Environmental Technologies (ITUT) Association was founded in Leipzig in 1997 as a joint initiative of the Federal government, the German economy and the Saxonian state legislature. Experiences from the redevelopment and structuring of the former German Democratic Republic, the existing know-how in engineering and planning due to the environmental protection level in Germany as well as the already developed environmental technology offered numerous foundations for possible solutions of similar problems in other regions of the world. By promoting the transfer of environmental technology and know-how, ITUT was trying to support the development towards a careful treatment of resources and the environment (ITUT 2001). Surprisingly, however, ITUT did not distinguish clearly between techniques and technologies. The exact translation of its German name *Internationales Transferzentrum für Umwelttechnik* into English would be "Centre for the International Transfer of Environmental Techniques". The same applied for the former "German-Japanese Co-operation Council for High-Technology and Environmental Technology (GJCC)" which terminated its work in 2002, where "technology" is replaced by the German expression of "techniques" (DJR 1999). Consequently, both institutes represent the tendency to use "technology" and "techniques" synonymously, although a clear distinction is well known. Moreover "Cleaner Production Germany"[30], an Internet portal of the German Environment Agency (UBA) developed to provide comprehensive information about the performance of German environmental technologies and services, fails to define ET. Consequently, one can easily assume that this could be a more strategic gambit for not limiting its activities.

The Japanese "International Centre for Environmental Technology Transfer (ICETT)", close to Yokkaichi[31], was established through the cooperation of industry, academia and government to serve as an organization affecting the smooth transfer of Japan's environmental conservation systems in order to con-

---

30 http://www.cleaner-production.de/en/ – 03 April 2010.
31 Yokkaichi (Mie Prefecture), about 330 km southwest from Tokyo, is one of the nation's most prominent petrochemical and industrial zones. Chemicals, like sulphur oxides, which exhausted from the chimneys, heavily polluted the air. Thus, air pollution was one of the four major cases of pollution in Japan. For detailed information see e.g. ICETT 1994a.

tribute to the conservation of the global environment and to the sustainable development of the world economy. But in contrast to ITUT, ICETT provides a comprehensive definition and description of ET:

> "The traditional means of combating pollution has been by means of end-of-pipe systems, i.e. treatment of wastes and polluting streams. This end-of-pipe approach, while still essential for many industries and for many technologies, should only be used as a last resort and Cleaner Production opportunities should be investigated first.
>
> Cleaner Technology is a manufacturing process which by its nature or intrinsically:
> - Reduces effluent and other waste production,
> - Maximises product quality,
> - Maximises raw materials and energy and any other input use.

Thus one technology is usually compared to some other technology or process. Cleaner Technology may be thought of a subset of Cleaner Production activities with a focus on the actual manufacturing process itself and considers the integration of better production systems to minimise environmental harm and maximise production efficiency from many or all inputs.

Clean Technology may be an impossible or difficult goal as it can be considered as the ultimate of the search for an inherently clean technology with no unwanted by-products, total use of inputs and full efficiency. On the other hand it may be used as a comparative term, and just be better than another technology. For example, considering membrane technology as inherently clean although even this technology produces waste streams." (ICETT 2001)

3.3.4 Categorization of Environmental Technology

Building on the above given definitions and descriptions ET can be separated into the following four categories:

1. Measuring Technologies on the Environment
   Tools, instruments, machines and complex systems, which measure and control or even harness the environment. One category of such technologies provides the necessary background information on departures from the natural balance. Another one is used to prevent harmful effects through environmental phenomena e.g. floods and shortage of water. In contrast to the following categories, the focus of this kind of ET is not on the minimization of anthropogenic impacts on the environment, but on the understanding of the environment and the containment of negative environmental impacts.

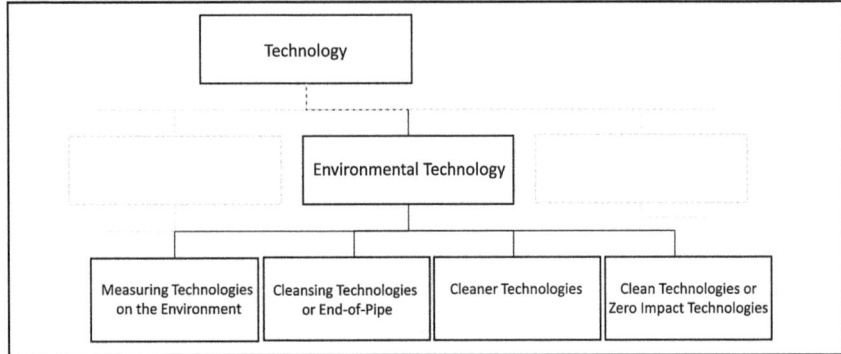

Figure 3: Categorization of Environmental Technology
(Source: Own Illustration)

2. Cleansing Technologies or End-of-Pipe Technologies
   Processes and materials that have been developed so as to minimize or neutralize harmful effects due to their use, without necessarily having to alter the original process. This cleansing technology is mainly based on end-of-pipe solutions, such as the instalment of exhaust catalyst and water filters (Almeida 1993: p. 5f). These technologies help in one direction to reduce the pollution of an environmental medium; however, they achieve the opposite in another direction through dilution, filters and recycling. Moreover, they determine a higher consumption of resources and energy, which leads to additional costs (Shen 1995: p. 236). These kinds of technologies have an additive or repairing function and constitute aftercare with a transformatory effect on emissions.

3. Cleaner Technologies
   Modifications to the production process may minimize or even eliminate the effects harmful to the environment, such as the introduction of control technology and changes in the types of raw or others materials (Cramer 1991: p. 461f). These integrated technologies are attempting to protect the environment through a holistic reflection of the entire product cycle (Brickwedde 1995: p. 261). Thus, the load is spread over several environmental media, which finally leads to reduction of environmental pollution through the preventive conception of these technologies (Förstner 1995: p. 488).

4. Clean Technologies or Zero Impact Technologies
   In contrast to Cleaner Technologies, Clean Technologies do simply not have any negative impact on the environment. This applies for example for certain methods of production in the chemical industry and in separating membranes

3 Definitions – The Key Terms

as well as the latest industrial conversion processes, based on special methods used in biotechnology (Bizri 1992: p. 32). Emissions register, however, on only a few important parameters as zero, with the others demonstrating acceptable emission values. Although the development and operation of clean or zero impact technologies seems Utopian from a thermodynamic perspective, they mark a "landing point". The maximal approximization to this point is the operational target. Nevertheless, some analysts believe that cleaner technology seems to be a more suitable term (Barnett 1993a: p. 9) since the fourth category of ET practically does not exist.

The ET of the first and second category has been more rapidly developed and adapted than those of the other categories. Especially measuring technologies are essential for understanding the interdependencies between the ecosphere and anthropogenic activities. Consequently, they can be considered as fundament or elementary ETs.

Nevertheless, the term "environmental technology" is usually used in conjunction with approaches to minimize the production of environmentally harmful substances. This target is met by the technologies of the second and third category. There is a common agreement that preventative measures are more economical than curative, meaning cleansing technologies (Jänicke 1988: p. 13). Likewise, the technologies of categories two and three are more widely employed than those of category four. This is easy to be explained by the need for radical transformation in most cases. The introduction of zero impact technologies is greatly hindered by linear thinking, not allowing for taking the necessary holistic view and a lack of investment in environmental protection through prevention, even in industrialized nations.

Consequently, the installation of cleaner and clean technology would meet the definition of sustainability by economic and ecological parameters. Higher investments for these technologies require more efforts to develop highly efficient and cheaper technologies – which may lead to models copying nature through integrated systems (Pauli 1996; Kühr & Széll 1997). Yet these medium- and long-term goals need the support of cleansing technologies to cope with urgent harmful emissions e.g. through highly polluted sewage in drinking water or highly contaminated exhaust fumes of incineration plants. Thus, a transnational environmental policy through the transfer of environmental technology has to take all four categories into account.

As a result the operational definition of ET could be:

"Environmentally Technologies (ET) contain four different categories: (i) measuring, (ii) cleansing, (iii) cleaner and (iv) clean technology. ET reduces pollution at

least in one environment medium ET is about continuous improvement of processes, products and services by the conservation of raw materials and energy and by the reduction of toxic substances, waste and emissions within the production cycle."
(Own definition, 2010)

This study will analyze not only individual techniques but also the entire system of environmental technologies including know-how, practices, goods and services and management. Focusing on the policy of Japan for the transfer of ET in Newly Industrializing Countries, this study will touch on the development of human potentials.

3.4 Newly Industrializing Countries

The term "Third World" is only a generic term used in political colloquial language. The official international nomenclature used by UN organizations is the term "developing countries". In the mid-1970s the theme of a challenge posed by the so-called Newly Industrializing Countries (NICs), Rapidly Industrializing Countries (RICs) or Take-Off Countries (TOC) emerged. And since the 1990s nation-states changing from centrally-planned economies to free-market models are referred to as Economies in Transition (EIT) or Transition Economies.

The traditional North-South division of labour, which was associated with the exchange of manufactured products for raw materials and primary commodities, was seriously called into question through the sharp increase in exports of manufactured goods by a number of developing countries (OECD 1988). They constituted the most visible result of the industrial take-off of a limited number of countries which had begun in the 1960s. In the eyes of the OECD, the increased participation of NICs in export markets for manufactured goods resulted, in broad terms, from the conjunction of two types of factors:

1. The choice of a development strategy based on the promotion of exports,
2. The establishment of foreign subsidiaries of multinational enterprises from the USA, Europe and Japan, whose production is largely export-oriented.

From the 1980s, and especially in the early and mid-1990s, there was growing international recognition of the rapid economic growth, structural change and industrialization of some nation-states. International organizations developed lists of those countries that surpassed the typical characteristics of developing countries and figuratively speaking were ready for "take-off". These nation-states are supported to a greater or lesser degree on exports from internally generated industrial production rather than on agricultural products or commodities. However, there were deviations in these lists, primarily because of various criteria and threshold values. The per-capita income, however, played a major role for all of them.

Although no official sanctioned category of NICs exist, most studies refer to Brazil, Mexico, China, India, Malaysia, Thailand, the Philippines and South Africa and Turkey as NICs (Mankiw 2007; Bozyk 2006; Guillén 2003; Waugh 2000a, b).[32] Few authors are also referring to Egypt, Indonesia and Russia due to differing results from the economic analysis performed (Bozyk 2006: p. 164).

Over the past decades the world has seen the stunning transformation of Japan[33], Hong Kong, Singapore, South Korea and Taiwan from impoverished developing countries into bustling, expanding economies. From 1965 till 2000 the latter four economies have quadrupled their share of world production and trade and quintupled their per-capita incomes (OECD 2001).

There has also been a tendency to see East Asia as much more of an economically integrated region than it actually is, and a corresponding tendency to see economic progress in the region as being similar in origin and nature. Terms such as "tigers", "minidragons" etc. have tended to perpetuate this perception. Although the indices of manufacturing sector growth were also impressive for three Southeast Highly Performing Asian Economies (HPAEs) until the late 1990s – namely Indonesia, Malaysia and Thailand – one can find important differences with other NICs. Besides having relatively lower economic growth, these NICs have also had relatively higher population growth, which means that the average living standards have risen more slowly. Furthermore in terms of the contribution of manufacturing to GDP, these NICs have also performed well, but not as well as the others.

Coming from a personal perspective, the author's consultancy with e.g. the German Gesellschaft für Technische Zusammenarbeit (GTZ) and the United Nations Environment Programme (UNEP) in the field of environmental technological transfer to Southeast Asia provided insights into the actual policy development and implementation. This is of relevance and contributed substantially to the final evaluation. Hence, although the term "NICs" refers to many more countries, this study will predominately focus on Japan's transnational environmental policies through the transfer of environmental technology to the NICs, analyzing Indonesia as a sample. Nonetheless, it is intended to refer descrip-

---

32 Until the early 1990s South Korea, Taiwan, Hong Kong and Singapore have been referred to as NICs. But since the United Nations Conference on Trade and Development (UNCTAD), OECD, World Bank and International Monetary Fund (IMF) are referring to the four nations as rather close to the socio-economic profile of industrialized countries. Taiwan declared itself already officially as industrialized (Almeida 1993).
33 Until the mid 1960s Japan was recipient of loans through The World Bank.

tions, analyses and conclusions to the majority of NICs, wherever possible and appropriate.

3.4.1 Rapid Economic Growth and the Environment

In its study *The East Asian Miracle* the World Bank argued in the early 1990s that Indonesia, Malaysia and Thailand provided the preferred models for emulation by other developing countries (World Bank 1993: p. 25). In comparison with the Northeast Asian experience of rapid growth and structural changes, those were accelerated by selective and judicious government intervention in the form of industrial policy. Thus, this World Bank study presents the success of these three NICs in the 1990s as proof that other developing countries can achieve rapid growth with minimal industrial policy and economic liberalization.

The study, which was commissioned on the insistence of the Japanese government to gain greater recognition and appreciation of the Japanese experiences with its rapid economic development after the Second World War,[34] pays surprisingly little attention to the fact that high growth has come with rapid expansion in e.g. energy and resource consumption, industrial and vehicular pollution, and hazardous- and ordinary waste generation.[35] The shift in economic structure from agriculture towards industry and services has – somewhat unexpectedly – been accompanied by a worsening rather than improvement in certain fields. Agriculture's ability to support a rapidly growing urban population has derived from crop intensification, involving the adoption of high-yielding varieties and the heavy application of fertilizers and pesticides. Consequently, the agriculture run-off contributes to e.g. water pollution through nitrates, phosphorous and toxic chemicals.

Rapid urban and industrial growth has brought serious pollution problems including:

---

34 This apparently came about after years of frustration with the free market conservatism and neo-classical economic orthodoxy, which has for a long time, dominated the thinking of the World Bank and policy recommendations, especially since the resurgence of the neo-liberal economic fundamentalism in the 1980s.

35 From 1975 to 1990, for example, per-capita energy consumption trebled in Korea and rose by a factor of 2.5 in Taiwan, Thailand and Indonesia. Private car petrol consumption grew almost 40 times in Korea and over 20 times in Taiwan. From 1981 to 1991 the total sulphur dioxide (SO_2) nearly trebled and nitrous oxide (NO_2) emissions rose by a factor of 2.5; in Taiwan and Thailand solid waste generation per-capita grew by over 50% (O'Connor 1994: p. 11).

3 Definitions – The Key Terms 37

- Air pollution from both stationary sources (power plants and industrial plants) and mobile sources (such as vehicles),
- Water pollution from oxygen depleting wastes (biological and chemical oxygen demand (BOD & COD), micro-organisms (faecal coliform) and hazardous substances (toxic chemicals, such as persistent organic pollutants and heavy metals),
- Growing volumes of municipal and industrial waste, including hazardous waste.

A traditional problem of the first phase of rapid urban and industrial growth is the lack of sanitary sewage disposal and the rise of household wastes through the rising urban population densities, overlapping with new growth-induced problems like industrial and automobile pollution. With increasing urbanization, industrialization and income levels, commercial energy demands tend to rise steeply, since high-income industrialized economies are relatively energy-intensive economies (The World Bank 1992; Kaya & Yokobori 1997).[36] However, since energy sources differ in their pollutant loadings, it is important to consider the country's energy mix. Coal is among the most heavily polluting source of electricity, notably in terms of suspended particulate matter (SPM) and nitrous oxide (NO_x). Coal and also oil can yield high sulphur dioxide (SO_2) emissions depending on their sulphur content. Nuclear power poses a variety of environmental risks, including fugitive emissions of radioactive gases, transport and disposal of radioactive wastes, not to mention possible accidents as in Chernobyl.[37] Although hydro-power is essentially clean, its development can cause immense damage to the natural resource base through e.g. flooding of the most fertile territories of a river basin.[38] Natural gas and biomass are by far the cleanest of the fossil fuels, though like the others their combustion generates substantial amounts of carbon dioxide (Barbier et. al. 1991: p. 122; Smith 1993: p.20ff).

The industrial sector itself has undergone major structural changes in all NICs. Following Jomo (1997) three phases of industrial development in NICs can be delineated:

---

36 In 1990, for example, the average energy consumption per-capita of high income OECD countries was almost four times that of middle-income economies and fifteen times that of low-income countries.
37 See the calendar of nuclear accidents at http://archive.greenpeace.org/comms/nukes/chernob/rep02.html –17 July 2010.
38 Some of many examples are the Angara Cascade in Russia and the Assuan Dam in Egypt.

- In the first phase growth comes largely from light industries like textiles and clothing and food and beverages, which rank relatively low in pollution intensity. Notably in Indonesia and Thailand resource-based industries loom large and cause many environmental problems. But the most serious industrial pollution problem of this phase is usually water pollution from BOD/COD releases.
- In the second phase certain heavily polluting industries take on importance, including such intermediates as ferrous and non-ferrous metals, petrochemicals, and non-metallic minerals e.g. cement and glass; air pollution is another serious problem though.
- In the third phase the leading growth sectors as e.g. electronic/electrical equipment, machinery and transport equipment tend to be of low to intermediate pollution intensity. The principal problems are from hazardous waste and resource consumption.

Consequently, along with the changes in the overall pollution intensity, the structural transformation of industry in NICs has brought about marked changes in the nature and composition of pollution streams and other impacts on the environment. Since different industrial sectors generate different types and volumes of pollution, industrial transformation can have significant environmental effects. To estimate those effects, one makes use of measures, tools and indicators of the relative pollution and environmental impact intensities of different industries in different countries, regions, urban areas or new projects. Such indices are calculated for different industrial sectors and products based on various, but sometimes complex data bases. Finally, they lead to the development to an indication of the environmental input of certain companies and industries.[39]

Rapid growth has created serious environmental pressures of at least two sorts:

1. By virtue of the sheer scale and, in many cases, geographic concentration of industry and related economic activities,
2. Because generations of environmental problems appear in rapid succession and thus partially overlap (O'Connor 1996: p. 5).

---

39 To estimate the pollution intensity of industries one makes use of e.g. so-called Pollution Intensity Indices, Toxic Chemical or Metallic Release Inventories. The calculations are either based on total releases or weighted sums of releases, with the weights being risk factors based on e.g. acute human toxicity, acute aquatic toxicity, and carcinogenicity.
Another rather popular tool is Life Cycle Assessments (LCA), which analyze the environmental aspects and potential impacts throughout a product life cycle (e.g. cradle-to-grave) from raw material acquisition through production, use and disposal.

According to the Environmental Kuznets Curve (EKC) at relatively low levels of income, the use of natural resources and/or the emission of wastes increase with income. But beyond some turning point, the use of the natural resources and/or the emission of wastes decline with income. Subsequent statistical analysis, however, showed that while the relationship may hold in a few cases, it could not be generalized across a wide range of resources and pollutants (Stern 2004: p. 1,419ff).

Nevertheless, there is common agreement that those NICs which had hoped simply to grow out of environment problems appear instead to be growing into them. It remains to be seen whether those which delayed decisive environmental measures will be penalized by slower growth in the future. What seems clear from the historical evidence is that those countries which addressed environmental problems early on did not pay a significant price in slower growth.

Additionally NICs have enjoyed certain advantages associated with being late but rapid industrializers. Among the most important are:

- The opportunity to learn from predecessors,
- The reduced uncertainty about absolute and relative magnitudes of environmental risks, which helps in prioritizing problems,
- A rapid uptake of technological innovations,
- The availability of a wider range of low-cost ET options which can be more closely matched to the country's environment.

In most NICs popular recognition and public concern over environmental problems have at least begun to appear on the political agenda. Although there are various environmental problems that need addressing, the national governments choose to concentrate their efforts on areas where the degree of political pressure for action is greatest. Also, with the East Asian currency and financial crisis of 1997/8, which radically transformed international perceptions and opinion on the East Asian experiences,[40] with earlier praise quickly changing into severe condemnation (Jomo 2001: p. 1), priority was given to solve this financial crisis.[41] Similarly, the present crisis illustrates the tendency to put certain long-term environmental initiatives back against short-time economic interests. The discussions and the final outcome of the United Nations Climate Change Conference in Copenhagen in December 2009 is one example highlighting this.

---

40 Especially the three second-tier Southeast Asian NICs – Indonesia, Malaysia and Thailand – as well as the Republic of Korea of the first generation of NICs has been most adversely affected.
41 For more information on this crisis see for example Pasuk & Baker 2000, Rasiah 2001, Jomo 2001, Montes 1998, McKibbin 1998.

Rapid economic growth and structural change, mainly associated with export-led industrialization in the Southeast Asian NICs, can generally be traced back to the mid-1980s. Then, devaluation of the currencies of all three Southeast HPAEs as well as selective deregulation of onerous rules helped to create attractive conditions for the relocation of production facilities in these countries and elsewhere in Southeast Asia and China. This was especially attractive for Japan, experiencing currency appreciations, tight labour markets higher production costs as well as increasing environmental awareness resulting in stricter legislation. Because of this the responsibility of Japan has risen for environmental influences in NICs, although these influences are basically of local or regional character.

4 Environmental Problems – A Challenge for Nation-States

4.1 Categorization of Environmental Problems

Environmental problems and disasters are a staple of the daily news. These problems are both, nature-made and man-made. Hence, they include earthquakes, hurricanes and floods, extreme droughts as well as depletion of the ozone layer and climate change. Some of these phenomena arise not as transnational consequence, while others are exclusively man-made. Nevertheless, all have negative consequences; they leave behind many casualties, lead to epidemics, homelessness, the destruction of vital resources or direct and/or indirect impacts on human health.

Quite often natural disasters must be considered as the starting point for a new vicious circle.[42] The swelling of the world's population and the growing interference in the environment leads increasingly to strong damages through comparatively small natural catastrophes, but also an increasing pressure on the limited natural resources and habitable space. Additionally, there is empirical evidence that the number of relatively large natural disasters are increasing; there is also consensus among scientists that the anthropogenic influence is growing. Protective measures against natural disasters like earthquakes that can result in the collapse of artefacts like buildings, bridges and highways are costly and do not present vested protection. Here it is appropriate to also point to the danger through damage to atomic power plants, pipelines, refineries and dams caused by earthquakes, especially when lack of money prevents the installation of safe techniques and required maintenance of early warning systems. These non-calculable sequels of natural catastrophes are comparable with the intervention in ecosystems (Böhret 1990: p. 40f). They also demonstrate environmental risks with partly global character that governments are required to meet through appropriate measures.

One can distinguish three different categories of natural environmental disasters:

1. Natural phenomena without serious impact on mankind such as the eruption of volcanoes,
2. Natural disasters like hurricanes whose negative consequences are not necessarily increased through anthropogenic influence,
3. Natural catastrophes directly or indirectly intensified through human activities with impacts on flora, fauna and human-beings. One example is

---

42 For more details see e.g. Maybury 1986.

the rampantly growing suburbs of mega-cities without the adherence to precautions against earthquakes.

In addition to the above described natural disasters, primarily man-made environmental problems with an impact on the animate and inanimate environment comprise the second group. Some natural incidents like special weather conditions e.g. in the case of the Indonesian forest fires in 1997 and 1998 covering large parts of Southeast Asia with smoke, might reinforce the harmful effects. Some of these problems can occur surprisingly as e.g. accidents in atomic power plants with the release of radioactivity or the emission of untreated sewage through an accident with heavy metals, hazardous chemicals and organic pollutants contaminating waters. Still another example is the growing greenhouse effects whose impacts are not experienced directly but rather progressively over a longer period of time. A specific example is the destruction of stratospheric ozone by chlorofluorocarbons (CFCs) whereby incremental impacts over decades lead to adverse impacts on ecological systems, animals and people, due to the increase in the amount of harmful infrared rays absorbed by the ozone layer (Rahmstorf 2007: p 29ff; Graßl 2007).

The above illustrates how natural disasters and man-made environmental problems can have impacts on the animate and inanimate environment. To complete the distinction between different categories of environmental problems, the third group does have immediate and exclusive consequences only for mankind through being the last link of the food chain. One example is the utilization of carcinogenic components within food production processes.

This categorization of environmental problems demonstrates the challenge states are confronted with in taking appropriate measures against them. In Germany they are laid down as a nation-state target since October 1994 namely in Article 20a of the German *Grundgesetz* (constitution) and in Japan's Basic Environment Law which was enacted in November 1993 as well as in specifying new ideas and various policies concerning the preservation of the environment.

4.2 Transnational Environmental Problems

Mankind produces consequences through its mere existence and action; thus, the amount of anthropogenic-caused effects and their interdependence with the natural system is incalculable. In this sense it is essential to extract special problem fields and to identify focal points as outstanding areas of discussions, with the aim to reduce the anthropogenic harmful influence on the environment. Follow-

4 Environmental Problems – A Challenge for Nation-States   43

ing the basic model of telic action,[43] the behaviour is steered by the goal aimed at (Ward 1907). Consequently, the environmental condition can be influenced through appropriate control sets and steering systems forming the basic model of environmental policy.

Transnational environmental problems are a relatively new area of public discussion, exemplifying the extent to which different nations are increasingly linked through globalization as they are sharing the same environment. In essence nine problem areas of concern have developed in transnational environmental relations with relatively identifiable transnational and global effects:[44]

1. Climate change,[45]
2. Loss of stratospheric ozone,
3. Loss of biodiversity,
4. Loss of forests,
5. Degradation of international waters,
6. Air pollution,
7. Exploitation and export of limited natural resources,

---

43   A model of thinking or a programme of action that assumes certain values as ends to be attained by deliberate, consciously planned conduct towards that end.
44   Air pollution must be considered as a clear transnational environment problem. For example the burning of lignite in East China and North of the Czech Republic caused acid rain at Japan's West Coast and in the Thuringian (Thüringer) Forest at in Southeast Germany, respectively. But as the World Bank has developed six problem areas of concern in global environmental relations, which are supplemented through "waste" by Simonis (1996), an analysis of the transnational environmental problems must also include the general topic "air pollution", as appropriate cleansing technology through e.g. filter systems are not yet exhaustively installed. Finally, the export of limited natural resources should also be added to this list. The waste of these resources often leads to enormous environmental impacts in the importing countries through backyard recycling practices with primitive techniques.
45   In June 1999 two islands of the Pacific nation of Kiribati were swallowed by the ocean. Tebua Tarawa and Abanuea were both uninhabited. The former was used by fishermen; the latter is known as "the beach which is long lasting".
Also in 1999 the South Pacific Regional Environment Programme explained that other islands of Kiribati and Tuvalu are in danger. Coasts were eroding in the Maldives and on the Marshall Islands where, according to the BBC, it would have costed US$100 million to build a seawall for just one atoll. The BBC quotes Maldives President Maumoon Abdul Gayoom in June 1999: "Sea-level rise is not a fashionable scientific hypothesis. It is a fact."
Since 1999 a number of islands disappeared. Last reported was the swallowing of the South Talpatti Island in the Bay of Bengal in March 2010.

8. Increase and export of hazardous wastes,
9. Increasing desertification.[46]

There is unanimous agreement that the problem of climate change is with global effect. For the large majority of analysts it appears appropriate to describe these trends with "globalization of environmental problems".

The German Advisory Council on Global Change (WBGU) defines global environmental problems as changes of the atmo-, hydro-, litho-, pedo- and biosphere, which are characterized through their origin in direct or indirect human activities, having impacts on the natural material flows, the aquatic and terrestrial symbiosis and also on the economy and society. To cope with global environmental problems international agreements are essential (WBGU 1993; Simonis 1996: p. 9).

All global environmental problems are in line with this definition. Following the WBGU global environmental problems could only be distinguished from those of local or regional character through the necessity of international measures against them. But not all environmental problems are of global dimension or even proving a globalizing tendency. Moreover, bilateral agreements between two nation-states or even multilateral settlements are not necessarily of global dimension, although they are international. This means that international environmental agreements are not necessarily dealing with global environmental changes. Consequently, the above-given characterization appears appropriate to operationally define

> Transnational environmental problems are changes of the atmo-, hydro-, litho-, pedo- and biosphere, which are characterised through their origin in direct or indirect human activities, having impacts on the natural material flows, the aquatic and terrestric symbiosis and also on the economy and society. To cope with transnational environmental problems international agreements are essential. (Own definition 2010)

Simultaneously, one could distinguish between:
- Transnational environmental problems of global character,
- Transnational environmental problems of local/regional character.

Hence, global environmental problems are transnational environmental problems of real global dimension.

---

46 The World Bank does only identify six problem areas (World Bank 1997: p. 5f). Despite the implementation of several cleaner technologies waste avoidance has not yet become a real social duty in most developed but also developing countries (Simonis 1996: p. 20).

4 Environmental Problems – A Challenge for Nation-States

A growing understanding of the origins, facets, developments and problem solving approaches regarding environmental problems does not only demonstrate the necessity for transnational action, but also provides the perspective for an increased efficiency through mutual-sustained measures. Although globalization brings along many transnational and global problems, it also offers new problem solving attempts through e.g. information exchange via the Internet and policies of rapprochement. These allow new perspectives for sustainable solutions and also a well-founded advice of experts thousands of kilometres away from the local/regional problem.

All nation-states around the world are affected by transnational environmental problems of global character as we have seen with climate change; responding to this danger calls for a common policy These two cases have in common that all nation-states involved are directly affected by the same problems. But this does not apply per se for transnational environmental policies. An exception represents the transfer of environmental technology from Japan or Germany to NICs in Southeast Asia.

The policies formulated and implemented in both countries aim at environmental problems of local/regional and global character. The problems these policies are responding to do not necessarily directly or indirectly affect Japan or Germany. Examples for the respective problem fields are sewage treatment in small towns or incineration plants for the growing mountains of waste. Certainly more and more international organizations and companies are becoming actors in coping with such regional problems and thus, influencing national approaches, especially when these seem to be insufficient and inappropriate. Thus, this study will take up the motives and targets of Japan in strengthening the transfer of environmental technology on the bilateral level through comparison with Germany. Additionally, it will analyze the input of international organizations and other non-governmental actors on transnational environmental policy.

5 Focus of the Study

This study focuses on Japan's transnational environmental policy through environmental technology transfer in the 1990s – on actors, coalitions and interactions across state boundaries. It treats and compares the reciprocal effects between:

- Transnational environmental policies and the effectiveness of national governance under the changed framework conditions through globalization,
- Environmental, technological and societal development,
- Eco-political development cooperation and national interests.

Wherever possible and appropriate, this study is going to make comparisons with Germany. This is to gain additional knowledge in answering the study's key questions and illustrate certain transferability to other nation-states.

The timely limitation of this study to mainly but not exclusively the 1990s is due to several reasons: First, Japan completed comprehensive environmental technology transfer projects at the end of the 1990s, allowing ex-post an internal and external assessment of their policies. Moreover, it puts the researcher in the position to represent the policy-cycle in its entirety and based on this draw conclusion for the verification or falsification of the theory on which this work is built. Second, the relative completion of this endeavour eases the research work in this rather complex policy field. No substantial adjustments of the respective policies were observed since the late 1990s, which would justify the shift of the time frame. Finally, much research for this study was performed in the 1990s during visiting fellowships at the Hitotsubashi University and the United Nations University (UNU) in Tokyo, Japan.

The analysis of Japan's transnational environmental relations does not explore them simply because "they exist". On the contrary, it is hoped to use this analysis to shed light on a number of empirical and normative questions that are directly related to the contemporary concerns of statesmen, scientists and other activists of this still rather new political field.

A glance at the table of contents of this study will reveal that it appears necessary to especially demonstrate that the transfer of environmental technologies to NICs is characterized through its high, multi-dimensional relevance. Chapter 34 of Agenda 21 highlights that when discussing the transfer of environmental technology, human resource development, local capacity building and gender-relevant aspects should also be addressed. Furthermore, these initiatives should be compatible with nationally determined socio-economic, cultural and environmental priorities.

Until very recently the scientific debate on problems involved with environmental technology transfer has been focused on two core aspects:
1. Research on developing countries,
2. Attempts to develop medium and long-term theories on development processes in the NICs.

Corresponding to the simplified representation of development cooperation as a relationship between the provider of aid and the recipient, the interest of political research in the past has been very much on the side of the recipient. Due to the dilemma facing discussions on development theory, the need to deliver nomological statements on the development process in the Third World, attention is being newly focused on the interests of those countries providing aid.

Table 2: ODA net performance (2008, 1998 & 1994)

	2008		1998		1994	
Country	Million US$	ODA/ GDP	Million US$	ODA/ GDP	Million US$	ODA/ GDP
Japan	9,579	0.19	10,64	0.28	13,239	0.29
USA	26,842	0.19	8,786	0.1	9,928	0.14
Germany	13,981	0.38	5,581	0.26	6,818	0.34
DAC (total)	121,483	0.31	51,888	0.24	59,152	0.30

(Sources: MOFA 2010 & OECD 1999: p. 62; 1998: p. A7f)

While Japan was the greatest provider of ODA worldwide in 1988 and 1989 and held this position again between 1991 and 2000, very few political analysts in the German speaking research world have devoted attention to Japan's development assistance (Kevenhörster 1993: p. 116-134; 1995: p. 5-21; 2006; May 1989: p. 40-59; Nuscheler 1990; 1994: p. 163-180; Rohde 2003; 1995a: p. 390-400; 1,995b: p. 601-612). Extensive research for information in Japanese on academic articles published in academic journals or university bulletins, or articles included in the National Diet Library's Japanese Periodicals Index Database, could not identify any studies on Germany's ODA.

Those English and German publications that do include development assistance as an element of foreign policy, trade and international economic policies, fail to describe the environmental dimension of development aid in any detail. Investigations dealing with the transfer of environmental technology (ET) as part of the Japan's and/or Germany's eco-political ODA remain to be seen.

Under the United Nations Conference on Trade and Development (UNCTAD) and through the financial support of the government of South Korea, a feasibility study examined the role of publicly-funded research in the transfer and diffusion of ET. This includes economic, legal and institutional issues, explores the potential of publicly-owned technologies and publicly-funded research to meet the demand for ESTs and examines specific mechanisms for enhancing their availability. Elements for an operational agenda, containing practical ways and means to promote, facilitate and finance the access to, and transfer of, publicly-funded ESTs to developing countries as well as economies in transition, are presented with a global outlook. In 2005 the 23rd session of the UNEP Governing Council approved the Bali Strategic Plan for Technology Support and Capacity Building. In 2007 the COP13 of the United Nations Framework Convention on Climate Change (UNFCCC) decided to launch a comprehensive process to enable the implementation of the Convention through a long-term cooperative action, the Bali Action Plan. These plans constitute an inter-governmentally agreed-upon approach to strengthen technology support and capacity building in developing countries, as well as countries with economies in transition. It seeks to strengthen the capacity of governments of developing countries and of NICs at all levels and provide systematic, targeted, long and short-term measures for technology support and capacity building. Another objective of the Plan is to promote, facilitate and finance access to and support for environmentally sound technologies and corresponding know-how. With these plans it is widely recognized that it is time to move on from Research & Development (R&D) to the implementation of innovative technologies. Sustainable Innovation, understood as the shift of sustainable technologies, products and services to the market, requires market creation concepts, a common global agenda and the transfer of the appropriate technologies.

In consequence UNEP and many other transnational/global players are putting a focus on the market potential of relevant technologies in the recipient countries. In addition to international organizations such as the OECD, UNCTAD, UNEP and UNDP, Japan, Germany as well as the USA, as lead nations in the field of environmental technologies, are playing an important role (BMU 2009). Key words such as "Green Gold" are making the round, whereas the focus of the nation-state is often only marginal on the sustainable effects of transferred technologies in solving the local and regional environmental problems. It is more on the enormous growth potentials of the export quota (see chapter 3.3.2) and hence the foreign trade aspects.

The Office of Technology Assessment of the US Congress analyzed environmental problems which less developed countries and newly industrializing economies are facing. Within this study existing markets for EST were identified, in particular on the commercial effects of the EST provided by other members of the Development Assistance Committee (DAC) on the American industry for environmental technology. Another paper by Moore and Miller describes

5 Focus of the Study 49

the efforts being made by Japan, Germany and the USA to ensure their economic future in the rapid growth market of EST (Moore 1994a, b). A recent OECD Working Paper analyzes the literature on policies for the development and transfer of eco-innovations (Popp 2009); the latest Environmental Technology Atlas for Germany authored by Roland Berger Strategy Consultants also commissioned and published by the Federal Ministry for the Environment focuses on Green Technologies made in Germany and the lead markets for environmental technologies (BMU 2009).

One has to bear in mind that there has hardly been any scientific discussions on the transfer of environmental technologies to NICs as part of transnational environmental policies. This is a result of the domination of two orientations of eco-political development cooperation: First, one understands eco-political development cooperation as part of development cooperation. Second, one focuses its principle that eco-political development cooperation is part of environmental policies' portfolio of tasks. In consequence social and economic aspects are falling behind. All in all the combination of quantitative and qualitative aspects has proved to be an obstacle in existing approaches to operationalize the paradigm "sustainability". The social and natural sciences substantially emulate each other, e.g. for research funding, hindering the necessary reflection of values and powers essential for the ecologic, social and economic development of societies (Flyvbjerg 2001: p. 53ff).

For the above reasons this study attempts to analyze Japan's policy for a clearly restricted field of work – the transfer of environmental technologies to NICs in the 1990s – as well as concerning the integration and balancing of social, economic and ecological dimensions, reflecting the values and powers of both nations. This is linked to the often repeated thesis that Japan's Official Development Assistance (ODA) is mainly part of its foreign trade policy (Kevenhörster 1995; May 1989: Nuscheler; Rohde 1995a). Meanwhile, in Japan and Germany the inflationary categorization of technology to be exported as environmentally-sound brings along the question whether all stand firm in a holistic environmental assessment or whether these attempts are to benefit from the growing markets worldwide.

The questions remaining constitute the theory/theories guiding this study, the methodological framework and the applied methods.

6 Theoretical and Methodological Approximations

6.1 Theoretical Framework

Karl *Popper* referred to theory as the net one casts in order to capture the world, to rationalize the world and to explain it (Popper 1976: p. 31). If this indeed applies, various nets are capturing different worlds leading to varying explanations. This has a consequence for the identification of causes and effects, of elements of permanence and elements of change, of criteria for validity and non-validity of statements.

> "Their observation acquires form through conceptual formulation, not from empirical reality (...)" (Rosenau 1990: p. 76).

To continue with Popper's metaphor, the following attempts to cast a net to assist with capturing Japan's and Germany's states' roles in transnational policy fields. For this purpose it appears necessary first to present and analyze the changing role of nation-states.

6.2 Changing Role of States

In the international state system, states use a specific set of rules to relate to each other (recognition of sovereignty and domain, agreements and treaties, representation, diplomacy and negotiations, balances of power up to and including war calculations). In contrast the arenas of markets and cultures are inherently more fluid, having rules which are less formal and do not correspond by necessity to particular national borders or to notions like sovereignty. Additionally John *Herz*, Jospeh S. *Nye*, Robert *Keohane* and many others already question the sovereignty of a state itself under the problem of defining national interests and what its defence requires under the conditions of the Cold War. In their opinion many of these uncertainties have their profound cause in the fundamental changes which have taken place in the changing nature of domestic politics in various states, advances in science and technology and thus also in the structure of international relations.

Their thesis reflects one central subject of political scientific discussions: The question regarding the role and influence of states. At present, especially taking the transnational environmental problems into account, states are in a dilemma: The efficiency of political regulations by a nation-state depends on the fulfilment of the congruence of state authority applying to an area, in which the people are living. This reflects the theory of three elements of international law, meaning that the area of interaction must not be much bigger than the area covered by political regulations. As long as economical activities, communication among the residents through telephone and letters, the creation of arts and culture and last but not least the pollution of the environment are carried out as far as possible within the framework of national border lines, nation-states are able to regulate those activities through appropriate steps (Zürn 1998a: p. 298f). But more and more incidents and impacts can be noticed globally with decreasing delay of time through such motors as market dynamics and new production and communication technologies. Global industrial concerns, a flood of information through technologies like the Internet difficult to control, transnational financial and economic links, increasing migration and proven impacts on the global climate through mankind are just some problems breaking the congruence of the nation-state. As a consequence the clear separation between national environments is eroded as the transnational and global impact of local influences emerges. This led to the coinage of the expression, *Our Global Neighbourhood* (Commission on Global Governance 1995). Moving from the national state to the transnational or global one, whereby the world becomes united to the extent that is regarded as one place and one global culture, poses a number of challenges for nation-states (Wild 1994; Robertson 1990).

Furthermore, it makes the dilemma clear: Modern states are confronted with a burden to make decisions, but decisions made do not necessarily have an effect on the problem without transnational harmonized action. This can finally lead to the development of new problem fields – for example the transfer of environmental harmful manufacturing to countries with comparable lax implementation of environmental regulation and thus to the migration of jobs and capital. The desire to maintain outdated policies can be interpreted as a) a lacking competence to solve problems or b) a lacking interest to solve problems. Factors such as public pressure and costs are playing a major role for the latter.

The political system is given the task to master the new conditions at the sight of changes in the system bridging a separation of space and time, immersing national economies in a sea of global flows and to cope with an increasing number of environmental problems. Additionally, there are more scientifically supported fears indicating that life might be irreversibly impaired in the near future, if certain global problems are not quickly and globally abolished.[47] Moreover, poorer countries raise environmental awareness and an associated ecological growth in criticism after expressing their opinion during the first UN environment conference 1972 in Stockholm that the "environmental gossip" is exclusively a problem of the countries in the North, while one must dedicate oneself in the South predominately to economic development (Weizsäcker 1997: p. 17).

---

47 See for example BMU 1994; Environment Agency 1995.

Modern states are heirs of the traditions of an authoritarian and benevolent state. As such states they are indebted to a political system for their inability to be equal to the challenges of social decisions. Furthermore, the logic resulting from these prerogatives is the lacking capability to give up collective decisions (Willke 1996: p. 11).

6.3 Academic Discussion

Put simply for a long time political science was conducting its research and the development of its apparatus of terms and theories so constrictively as if there would only be a policy among nation-states at the international level. As a consequence the research and subsequent explanation of processes between states was de-coupled from those within states over a long period of time. Nevertheless, it was a long time ago that one realized that effort for a theory of international policy requires an analysis of the processes within nation-state units and between those (Alger 1963). This also applies for an analysis of transnational environmental policies, which is the focus of this study.

In this context Karl *Kaiser* describes the structures and processes of research on international policies until the 1960s, treating the structures and processes within nation-state units as so-called "black boxes" (Kaiser 1969: p. 81). These "black box" nation-states were existing but their contents were not taken into account in studies on theoretical aspects of international policies. In Kaiser's eyes this also applied to research on domestic policies and governmental systems and the treatment of international policies. Today, some scientists are going so far to claim that national policies are rather insignificant for the analysis of international policies and structures because national policies are continuously losing importance due to globalization.

The boundaries resulting out of the division between national and international policies are illustrated by the transfer of environmental technologies to Newly Industrializing Countries: Here at least one political system each participates in the decision-making on the side of a Newly Industrializing Country and the (Post) Industrialized State in very specific fields. In addition there is a multitude of non-state actors; their participation constitutes from the joint planning, the joint implementation of programmes until possibly the establishment of joint institutions with personnel of both systems. It becomes obvious that actors from outside the political system both on the side of the environmental technology transferring country and the environmental technology recipient country are playing a major role. Actions based upon this are in turn determined by a) the structures of domestic policies, b) the objectives of foreign policies and c) the structures and agreements of the international systems as such. This results in complex interactions of processes of societal actors. These interactions are simultaneously obvious in at least two nation-state systems on the transnational level and governments, in which governmental governance represents only a tiny fraction of the entire phenomenon.

A theoretical and methodological approach is therefore vital in characterizing structures of decision-making and reflecting the relationship patterns and relevant values. At the same time this approach shall assist in presenting the roles of actors, the nation-state and national governments under these rapidly changing framework conditions of this policy field. Thus, it appears, above all, necessary to give an overview on theoretical approaches in the following, which seem to be relevant to uncover answers to the core-questions of this study. Accordingly, it also seems to be essential for the realistic assessment of the transnational environmental policies of states to briefly reflect the political scientific debate in this connection, primarily because many authors coin the expression of e.g. failure of state (Jänicke 1979; Mayntz 1979; Willke 1983; Recktenwald 1978; Luhmann 1988b; Beck 1986; von Beyme 1990).

6.3.1 Neo-Realists and Globalists

Since the 1970s social scientific discussions arose on the continuously increasing undermining of border lines through the growing importance of political, economical and social transnational activities (Messner 1998: p. 14).[48] In the 1970s the question was mainly how to overcome the economic crisis after the decades of rapid economic growth and how to overcome the negative consequences of the oil-shock. For some international theorists there was empirical evidence of state decline.[49] However, the ecological and social question was added to the economic problems in the 1990s. With the United Nations Conference on Environment and Development (UNCED) in Rio de Janeiro, Brazil in 1992, it became obvious that nation-states, international institutes and organizations as well as transnational companies lost considerable sovereign power to define the political agenda at the sight of e.g. financial instability, a growing drug trafficking and arms trade, an increasing gap between rich and poor, a continuous increase of unemployment despite national labour market policies and an dramatic ecocide.

---

48 It is impossible to summarize the burgeoning literature devoted to the idea of globalization. Some standard English references are: Albrow 1998, Jones 1995, Hirst & Thompson 1996, Robertson 1992, Rosenau & Czempiel 1992, Waters 1995.
49 See for example Keohane, Robert O. & Joseph S. Nye (1977): Power and Interdependence: World Politics in Transition, Boston, p. 7f.

The environmental policy discussions picked up the thread of diagnosis in the 1970s indicating that the ability of modern industrialized states is conspicuous and systematically affected (Jänicke 1993) and a number of analyses point to the relative weakening of state sovereignty. In 1979 even the German Chancellor Helmut *Schmidt* described the governments' possibilities of influence in modern industrial mass democracies as much less than imparted in school books (Spiegel 15 January 1979). Nevertheless the same Chancellor explained six years later the G7-summits[50] as informal and private meeting of those, having something to say (Putnam & Bayne 1985: 23). With reference to Schmidt's statement in 1979, this could be interpreted as an increase of power, which would stand in sharp contrast to the identical theories of many political scientists. Terms like "symbolic and ritual politics" (Edelmann 1976: p. 189), "decisionless decisions and non-decisions" (Bachrach & Baatz 1977), "looseness of control" (Lindblom 1980: p. 124), "social controllability problem" (Mayntz 1987: p. 106) and "passing state" (Ronge 1980) reflect these developments.

The self endangerment of mankind through ecological problems is often stated as a proof for the failure of governance, political control and the obstruction of communication among the parts of the system. The autopoietic[51] system theory with its scepticism of a state's ability to control was developed (von Beyme 1996: p. 14) as well as Luhmann's obvious "social control scepticism" and Willke's "tragedy and irony of state" can be mentioned as some outstanding examples.

Above all the claim is made that the sovereignty and autonomy of nation-states has been radically reduced.[52] At the sight of Schmidt's statement a contro-

---

50   Summit of Canada, France, Germany, Great Britain, Italy, Japan, USA. Henry Kissinger, the then US Secretary of State, first proposed the idea of a multilateral economic summit meeting in 1971. A few years later, Valéry Giscard d' Estaing, a former member of the Library Group who became president of France in 1974, invited his colleagues from the above-mentioned countries with the exception of Canada to the first economic summit held at the Château de Rambouillet in 1998, the place of the unfortunate peace talks between Serbs and Kosovo Albanians in November 1975. Canada joined a year later in Puerto Rico to form with the other six countries the Group of Seven (G7).
Already in 1967 Great Britain was host to a meeting of five Finance Ministers who discussed monetary flows and fluctuations in interest rates. Later, the so-called Library Group of Finance Ministers recreated these informal exchanges and first met in the While House Library in April 1973 to discuss international monetary negotiations.

51   A system can be described as autopoetic if its functions are addressed to renew itself; e.g. cells and their interplay of anabolic (constructive) and katabolic (destructive) processes.

52   See for example Richter 1992; Altvater 1998; Messner 1998.

versy between two different points of view must be assumed. One exclaims a decline of national sovereignty and efficiency and still the other claims a shift of the disposition to a transnational level. This has led to a flaring up of the discussion of so-called realists and globalists, which was predominately lively in the 1960s and 1970s. The phenomenon and policies of the integration, supported by the quantitative and qualitative enormous ascent of non-governmental actors, which are both bridging and undermining the sovereignty of nation-states, was rooted in their competence (Meyers 1991: p. 292). With this the summed-up premises of the neo-realists were that states are the only relevant actors and other international actors, in turn, are only tools, agents or simply receiving orders of states. In contrast to this, the globalists believe that states are not the only key-actors. In their eyes certain international transactions and results can only sustain due to the motives and action of transnational governmental or even non-governmental organizations. Long-lasting or ad hoc transnational coalitions, which appear irrelevant from a state-centric point of view, are gaining importance. Moreover, this discussion is fed by the growing perception of interdependencies, not only through global economic networks, but also through e.g. an internationalization of investments and production i.e. the capital market, boundless communication flows and the globalization of consumption-musters. There is also an increased perception of the global dimensions of security threats, poverty development, violations of human rights and environmental destruction; out of these interdependencies arise dependency with differing effects for those involved. Nevertheless, it increases the vulnerability through external developments and at the same time minimizes the possibilities for own problem solving attempts. There is consensus that coordinated activities beyond the borders of nation-states are necessary. But does it go so far that the nation-states are becoming an anachronism? Or do they remain the main actors? It is not in the interest of this study to look for a solution how an e.g. "governance without a world government" (Kohler-Koch 1993), a "cooperation under anarchy" (Axelrod & Keohane 1986) or "governance without government" (Rosenau & Czempiel 1992) might be possible. Rather, this work intends to find out how the Japanese state defines and implements its respective tasks in conjunction with the transfer of environmental technologies to NICs as part of its transnational environmental policies. Moreover, this study analyzes existing policies with the intent to identify evidences that a) nation-states are not the only actors in this policy-field, b) in how far transnational actors are influencing the formulation and implementation of the policies and c) if these indications give reason to support the premises of the neo-realists and globalists.

The issue, as it remains, is to find out which theoretical approach and which specific methodologies appear promising for this study. This will be the chief focus in the following sub-chapter.

6.3.2 Specific Approaches

Birgit Mahnkopf and Elmar Altvater examine obvious ecological and social boundaries for the accumulation and expansion of capital despite all rhetoric of free trade and globalization through, as they call it, "external forces of circumstances" instead of so-called self-assured politically-set limits. These forces of circumstances present themselves therefore in the argumentation of the two internally as crisis, which can neither be avoided within the economic logic of borderless expansion and with available tools of nation-states nor treated in an appropriate way (Altvater & Mahnkopf 1997: p. 376). They demonstrate that nation-states lose a part of their political control over the territory against global powers due to the economic approaches to set the value of a region, which they understand as an "exploitable mine" and not as habitat. Resources identified to be valuable, which also includes know-how, will be removed from the region through transportation and communication media and finally transformed into goods; this means a monetarization of resources on the global market (Ibid.: p. 380). Thus, the process of setting into value has both negative and positive effects on other regions by the taking of resources. In the last step of this model of setting value from Mahnkopf & Altvater, decision makers of the region become subject of the rationality of cash and capital so that all action is controlled by striving for profitability and profit through entering the global currency market; territorially bound statehood is lost because decisions follow an economical and not a political logic. Nevertheless, nation-states do not disappear, but rather they transfer to operate as special managers of national economies competing with each other. National areas can be seen as negative barriers hindering the liberty of acting, the capital flows and which obstruct services and the migration of its people. The perforation of border lines through economic globalization finally leads to the drawing of limits resultant from an ecological crisis. This developed from the tendency to subject politics and societies under the now and then destructive law of markets (Ibid.: p. 503).

Mahnkopf & Altvater come in their analysis to the result that the world functions according to other rules than under the design of "global governance", since powerful political and economic interests insert themselves only into a network of governance, as they profit from it. For a network of "global governance", it is unclear how it could deal with clashes of interests in economics, society and politics (Ibid.: p. 556). Leaving globalization to itself which would in turn also leave the big economic powers like transnational companies and global playing banks to themselves would ultimately lead to a global, social and ecological collapse. According to Altvater & Mahnkopf political boundaries should be set to put a stop to e.g. the exodus of capital and to keep the voice of the democratic public as loud as possible. This taming of globalization must take place outside the geopolitical sphere of nation-states and outside the sphere of nodalized[53] networks through a global civil society (Ibid.: 590), taking place "on the spot", meaning national societies, international networks, NGOs and international organizations. Consequently, new players (polity) will emerge, utilizing new problem solving approaches (politics) addressing new problem fields (policies) on a global level.

From Albrow's point of view the above-mentioned practical experiences of an epochal change give evidence to something new which cannot and should not be classified as part of modernity, but as the beginning of a new and global period. In this sense Albrow explains that the fundamentals of trade and social organization have changed, both for the individuals and for groups, as at least the following five aspects exceed the prerequisites of modernity:

1. Global environmental impact through anthropogenic activities,
2. The loss of every kind of safety through weapons and techniques with global power for destruction,
3. Communication technologies spanning the world,
4. The emergence of a global economy,
5. The values and convictions of all people are deduced from a global context (Albrow 1998: p. 14).

Out of the beginning of a new global period, a new political order crystallized; policy is far behind the above-explained fields. As politicians were highly profiting from the unity of state and society during modernity, these politicians have had a hard time accepting the transformations of the conditions through the uncoupling of culture, community and relations and their break-out of systems of nation-states (Ibid.: p. 253). The increase in transnational ties unmasks nation-states as transient structures, trying to face the challenges of the new period with inappropriate equipment, rhetoric and solution patterns of the past. Albrow as well as e.g. Geoff *Mulgan* 1994 and Paul *Hirst* 1993 criticize a discussion focusing only on the effects of governments' future economic policies irrespective of all empirical verifiable trends through globalization among other things on the

---

53 Nodalization means the development of a global economic sphere with an orientation through statistics of global capital flows, if available and reliable, instead of natural and political boundaries.

environment (Ibid.: p. 264). This indicates a comprehensive transformation, changing the overall conceptual framework, as increasing activities of states do not necessarily show a growth in power. Governments are losing their influence on the destiny of their people within a global polycentric network, setting rules in many places and being administered in just as many centres, with citizens living their lives as independent individuals with national hopes often spanning across states' border lines. With the beginning of the global period, linkage between nation and state is cut and more and more transnational activities are moving into the centre.[54] Following Albrow's argumentation that the decline of nation-states is already accompanied by the emergence of a global state, growing out of the combined efforts to cope with the immense environmental impacts through technological innovations, to be successful in a worldwide application of human rights and represented through the common fear of a nuclear catastrophe[55]. However, a global state will never function as a nation-state just on a global level; its circumstances of origin are not connected with aspects of territoriality and nationalism. Public pressure on national governments point out through the activities of millions of people fighting against global topics such as whaling, female circumcision and the water quality of the Mekong River in Asia or the Elbe River in Europe.

Instead of using the term "globalization" Michael *Zürn* pleads to prefer "social denationalization". He takes the growing social interactions into account, which are increasingly crossing the borders of nation-states; nevertheless new boundaries develop (Zürn 1998: p. 16). His use of a different term signal the start of a period in history distinct from the recent period which has been dominated primarily by nation-states acting in concert through the creation of an international economic system for trade and payments. That era has run its course and has given way to a period which is characterized rather through the increase of intensity and range of transboundary processes of exchange or production within the economy, environment, violence, mobility, communication and culture (Ibid.: p. 76).

Four normative goods reflecting the desires and values of the majority of people living in the OECD countries represent the aims of governance in complex societies with ambitious objectives having a high degree of independence competing with other societies. They can also be regarded as functional goods,

---

54 This cut between nation and state can also result in the unleashing of new nationalism, but on the other hand it supports a clear separation of the idea of states and the moods of governments (Albrow 1998: p. 267).

55 Especially after the worst nuclear accident in Japan happened in Tokaimura/ in September 1999, followed by the recommendation of the Japanese government to the people of the Kanto region to remain indoors.

6 Theoretical and Methodological Approximations

indicating a crisis of the political system if one or several of these targets remains unachieved for a certain period (Ibid.: p. 41). These "aims of governance" are:

- The guarantee of internal and external security,
- The creation of a symbolic reference system which offers the framework for development of a civil collective identity
- The acceptance that political decisions are approvable,
- The support of economic growth and the containment of social inequality through measures ensuring a broad material affluence.

In the course of social denationalization, the congruence of political regulations and the attainability of social interactions become a factor for determining the efficiency of governance, as national political institutions of the post-war era are attacked through e.g. proving weak in responding to the climate change issue and the disappearance of Japan's coastal forest due to acid rain from lignite burning in China in the 1990s. De facto nation-states are successfully cooperating to solve ecological risks but they have to meet certain requirements (Ibid.: p. 180). Thus Zürn exhibits through his analysis of international environmental regimes that positive regulations between states are possible and currently existing, although the respective countries differ in their economic productivity.

Both the theories of the decreasing effectiveness of national governing and the theory of a growing autonomy of the political class through denationalization are justified. National governments strive for recovery of effectiveness through international institutions; there are clear signs that those institutes evade from national social influence and control, but simultaneously prejudge national decisions. Under the specification that states traditionally attained the acknowledgment of the right to self-determination, both the national society and the national state would be the losers. A permanent legitimation control of states by the evaluation of international institutions in e.g. the economics and environmental areas, by observer missions during general elections and by the control of adherence of human rights through e.g. a UN court of justice, the role of the nation-state must be redefined. A shift of significance becomes obvious. Nevertheless, special areas remain existing in which the social denationalization does not have influence on the effectiveness of national policy. Areas are added like the constitutional state function in OECD countries, which must be evaluated as positive. Additionally, there are no signs for international institutions that could better achieve the targets of national governance (Ibid.: p. 333). Within this new statehood governance is held through an increasing cooperation of different decision levels and an executive body which will no longer be fully functioning without each other.

Starting from the crisis theory Jänicke explains an increase of environmental problems with the systematic problems of the industrial system. Its costs devolve upon the general public and additionally being a source of growth – also for the endangering of the environment (Jänicke 1979: p. 7). He highlights an extensive monopolization of the problem and target definitions through an industrial-bureaucratic symbiosis which results in fighting symptoms and a focus on costs. Consequently, preventive measures are neglected, although such a problem solving modus will be more costly in the long-term (Ibid.: 13). Jänicke sees the role of the state predominately in bureaucratic forms ensuring infrastructural and regulative achievements for the economic system and in the management of external effects of economic processes (Jänicke 1988: p. 18f). This leads to an increase in power of the state's bureaucracy and a decrease in sovereignty of political instances. The result is an economic-technocratic strategy of fighting symptoms which basically renounces preventive political interventions (Ibid.: p. 36; Weidner 1996: p. 72). According to Jänicke a comprehensive innovations' process of the industrial system could lead a structural change resulting in social innovations with a consensual planning of framework (Jänicke 1986: p. 166).

Long before Altvater & Mahnkopf, Albrow, Zürn and Jänicke made their contributions to the theoretical discussion, Georg *Jellinek* determined the theory of state of the 20th century and did, however, influence Max *Weber* (Weber 1990: p. 149). Jellinek defines political science as

"(...) the science of reaching certain national purposes and therefore the analysis of national appearances (...), which at the same time supply the critical measure for the evaluation of the national statuses and conditions"[56] (Jellinek 1922: p. 13).

Following Jellinek one can consequently define politics as practical state science. His strategy combines policy analysis with a political history of ideas. Thus, national purposes can only be determined through an analysis of national institutions and functions through a contemplation of the historical development and different existing points of view (Ibid.: p. 255; Ibid.: p. 259), according to the maxim of the historical school. Institutions are continuously changing and therefore adjusting new objectives; duties which require a central guidance should be performed by the state. The administration, preparing the legislation,

---

56 Translated by the author from the German original, Die Lehre von der Erreichung bestimmter staatlicher Zwecke und daher die Betrachtung staatlicher Erscheinungen (...), die zugleich den kritischen Maßstab für die Beurteilung der staatlichen Zustände und Verhältnisse liefern.

supporting the judicial work and ensuring the execution of the judiciary fulfils a basic function. As legislation is occupying increasingly larger parts of society, its power is also continuously increasing (Anter 1998: p. 510f). After Lorenz *von Stein*[57] and his friend Max *Weber*, Georg Jellinek's work must be considered as another trailblazing piece for political scientific sociology which still exists today and has a power of persuasion so that its positions and terms are still of relevance in current discussions (Ibid.: p. 523). Jellinek understands the nation-state as an institution, which is characterized through its functions and thus through its purposes.

After Jellinek's death the question regarding the state's purpose has moved back into the centre of scientific discussions in the 1960s and 1970s, although Hermann Heller's science of state is linked up with this question, a question he calls the most fundamental problem of state theory (Heller 1934: p. 200). Nevertheless, Heller's work was recognized as the crone of papers on the theory of state in Japan in 1935 (Dehnhard 1996: p. 25).

In the 1970s the question regarding the state's purpose was brushed aside through an alliance of positivistic and materialistic realists. At the sight of big political challenges through the above-analyzed various dimensions of globalization of the youngest past and the present, questions regarding the duties and goals of the state are being discussed again. Those analyses, as presented above, are dominated through an empirical-based view, which does not lay down the state's purpose unanimously, but follows, however, Jellinek's prints.

Albrecht *Dehnhard* developed his own theoretical model in the tradition of Jellinek and Heller emphasizing an interdisciplinary and multidimensional approach to create a holistic view in responding to the pressing problems of mankind. In his eyes an efficient and useful state theory had to identify all relevant factors with influence on the state's constitution (Ibid.: p. 6). Accordingly, Dehnhard is in line with Kaiser's general argumentation described above. In their eyes the identification of theories should integrate such aspects which are not directly linked to the respective policy field, but having an impact on such. The governance of a state and its actors is not only essential for a realistic presentation, but also to keep the freedom of action at the sight of unstoppable bureaucratization. In the eyes of Heller, there is a dialect between organization

---

57 Lorenz von Stein (1815-1890) is one of the great German social scientists of the 19th century, spending most of his life teaching at the University of Vienna. He influenced the practice of public finance but is perhaps best known for his sociological ideas, set forth in the third edition of his history of the social movement in France. He outlined an economic interpretation of history that included concepts of the proletariat and of class struggle, showing quite some similarities with Karl Marx, though von Stein's influence on Marx remains unclear (Mutius 1992).

and freedom; in his mind modern societies must organize their freedom (Heller 1992: p. 389).

Dehnhard argues that the description of reality has to take the cultural traditions into account. He demands the reflection on and integration of the personal characteristics as orientation, motivation, qualification of those human-beings involved in the state beside its structure, institutions etc. He pays special attention to learning and designs an action-related approach (Ibid.: p. 8). Although states are confronted with a changing framework, they cannot be described as stripped of all political power. Transnational and global environmental problems require a transnational and/or international policy management which is pursued by states. Supranational organizations like the United Nations (UN) are prevailingly depending from the power and actions of its Member States. According to Dehnhard states are losing power and influence at the international level, but simultaneously getting power and influence back at other levels. Thus, on balance they are gaining power (Ibid.: p. 11) and they still remain key-actors in international policies. To work out a comprehensive balance sheet of the current governance one has to examine the results of national activities, transnational approaches and the achievements of existing national and international institutions.

Social scientists differ in their assessment about the future of the nation-state and politics and strategies capable of bearing policy formulation under this changing framework of transnationalization and globalization. However, at least two major deficits make clear that these targets are hard to achieve under the current governing system:

1. An increasing imbalance of the distribution of affluence around the world,
2. Increasing serious global environmental problems.

Out of this panorama one can work out four far-reaching viewpoints:

1. The leave-taking of the nation-state through its loss of importance in responding to the pressing problems,
2. The lack of alternatives and the weakness of supranational structures keep the nation-state at the centre of politics, although it loses its ability to govern,
3. A growing multilateralism towards the strengthening of transnational cooperation, which will keep the nation-state in the position to be the main political player,
4. The transformation of politics into the design of global governance, forwarding responsibility to the international level after early diagnosis, the

development of solutions and the implementation of politics on local, regional and national levels.

All of the above have in common that the nature of world politics is changing. Furthermore, as explained above, there is consensus that environmental problems are intricately linked with economic and social relationships. Taking these four viewpoints into consideration, this work will carefully examine, through the analysis of Japan's transnational environmental policies and a certain comparison with Germany, if and how both states were responding to the changing framework in the formulation and implementation of such a multidisciplinary policy field of environmental technology transfer to Newly Industrializing Countries, especially in the 1990s.

As long as modern industrial nations are organized as democracies with a parliamentary system, the state and its machinery should play the key-role to solve life-threatening problems for society, even if under the complex conditions of modern industrial societies modified functions are appearing. On the other hand, there are rather radical discourses moving beyond international to global politics, but assume that globalization diminishes the state element of "governance" as shown above. Perhaps they are basing their assessment on the classic definition of the modern state which appears inappropriate under the changing framework. While this literature maintains that governance now involves more than the nation-state, it implies also that this should lead to replace a state-perspective with a perspective of governance. Max *Weber* specified that

> "A compulsory political organisation with continuous operations will be called a "state" insofar as its administrative staff successfully upholds the claim to the monopoly of the legitimate use of physical force in the enforcement of its order."[58]

Following Weber, Anthony *Giddens* defines the modern nation-state as a "bordered power container". Yet many states today are small, weak, with problematic national coherence and above all only limited autonomy in any sense. With the demise of the imperial European state of the last centuries, the nation-state has become a more or less universal political form, spreading first to the rest of Europe, then what became known as the "Third World", and finally to the remains of the Soviet Union. But with becoming more universal, nation-states have also been shorn of the key characteristics of autonomous state power. Consequently, most nation-states are no longer autonomous classic nation-states as characterized in Weber's and Gidden's definition. Moreover, with the end of the Cold War, especially the Western states, including Japan, are becoming a global centre and have been able to utilize legitimate global institutions, notably the

---

58 Quoted in Mann (1986): p. 55.

United Nations, to underwrite their own global projection of power. Western leaders, including the Japanese, have developed a rhetoric of global responsibility but there is still a great reluctance to commit real resources or thought to the development of global institutions or to form a global social change, which might offer greater stability.

Whether state institutes are able to carry out institutional and organizational structural reforms in order to make appropriate transformations with regard to the problems, this appears extraordinarily difficult in view of high-grade organized national and transnational organizations and institutional restrictions (Weidner 1996: p. 63ff). Nevertheless, representatives of the autopoietic system theory such as Niklas *Luhmann* and the institutional or action-theoretical and control-theoretical approach of e.g. Fritz W. *Scharpf* also do not exclude the control ability of political systems theoretically. Thus this analysis tries to show that it might be erroneous to simply counterpose globalization to the state, as many sterile debates in social sciences are doing. The author argues that globalization does not necessarily undermine the state, but includes the transformation of state forms; it is both predicated on and produces such transformations (Shaw 1997: p. 498). Furthermore, the transfer of environmental technology into Newly Industrializing Countries proceeds on the assumption of the existent, but probably reduced capacity of the nation-state to act. Special attention should be given to the actors within this study and especially how they were influencing Japan's transnational environmental policies. It will also be important to analyze how far they were influenced from other-non national actors and which were relevant factors for policy-formulation and the implementation process. In the light of Jellinek's arguments, this does not only require a policy-analysis including the review of the history and the context in which policies were developed, but a holistic multi-dimensional approach – the basis of Dehnhard's theory. This study attempts to illustrate the transfer of environmental technologies and with reference to the already presented approaches if and how nation-states were gaining power in international circles and therefore support Dehnhard's neo-realistic theory.

7 Methodical Approach

After developing a concept of content and thus an operating theoretical approach or model and setting a methodological framework for this study, the question now is, which methodical approach appears appropriate for the research object (Windhoff-Héritier 1987: p. 117f.)?

Policy is per definition "the intentional behaviour of actors, having an interest in attaining a certain target". These interests differ between actors because of their varying responses to the changing framework conditions and threats. Possibilities also vary due to differences in perceptions and preferences; moreover, these perceptions and preferences are influenced by the respective institutional context in which they are interacting (Scharpf 2000: p. 74).

For a theoretical approximation based on the Dehnhard's neo-realistic theory, a combination with a methodological approach appears meaningful for this study that does not concentrate predominately on the analysis of the problems and the search for possible effective solutions against them. Previous expositions on transnational environmental problems and the new challenges for states aimed at analyzing and explaining the emergence and the reasons for these problems. They did indeed demonstrate that there is already common understanding, that on the basis of the paradigm of sustainable development the transfer of environmental technology to newly industrializing countries appears as an effective solution with possible effects on the original problem and the political environment as such.

The methodological guidelines for reformed social science through practising phronesis[59] as put forward by Bent *Flyvbjerg* (Flyvbjerg 2001: p. 129ff) provide a certain direction. Flyvbjerg argues that instead of trying to emulate the natural sciences, the social sciences should be practised as phronesis and to explicitly include considerations of power.[60] Phronetic social sciences explore his-

---

[59] Phronēsis in Aristotle's "Nicomachean Ethics" is the virtue of practical thought. Aristotle distinguishes between two intellectual virtues: sophia and phronesis. Sophia (usually translated "wisdom") is the ability to think well about the nature of the world, to discern why the world is the way it is (this is sometimes equated with science); sophia involves deliberation concerning universal truths. Phronesis is the capability to consider the mode of action in order to deliver change, especially to enhance the quality of life. Aristotle says that phronesis is not simply a skill, however, as it involves not only the ability to decide how to achieve a certain end, but also the ability to reflect upon and determine that end (http://en.wikipedia.org/wiki/Phronesis – 17 July 2010).

[60] Flyvbjerg does not enter the natural versus social sciences debate but rather points out that both methods have strengths and weaknesses: "…the social sciences are strongest where the natural sciences are weakest: just as the social sciences have not contributed

toric circumstances and current practices to find avenues to praxis; hence, it takes the necessary holistic approach. Richard *Rorty* summarizes the principal concern of phronetic social sciences as

> "[Political Situations] get clarified by detailed stories about who's doing what to whom." (Rorty 1994: p. 14)

Accordingly, Flyvbjerg's phronetic social science focuses on four value-rational questions:

1. Where are we going?
2. Who gains and who loses by which mechanisms of power?
3. Is this development desirable?
4. What should we do about it?

Already in 1984 Udo-Ernst *Simonis* called for close cooperation between technicians and economists for the development of social and environmentally-sound techniques and their respective implementation, emphasizing the need for the necessary stakeholder involvement in this discourse (Simonis 1984: p. 9). Latently, the Agenda 21, which is considered as a dynamic programme towards sustainable development, provides instruments and tools towards the above illustrated aim and answers the four questions[61]; but so does the Plan for Implementation, the main outcome document negotiated by all states participating in the World Summit for Sustainable Development (WSSD) in Johannesburg, South Africa in 2002.[62]

7.1 Actors Analysis

This study does not focus on problem and solution-orientated research, because the specific contribution of political science and political sociology must be questioned in an area like transnational environmental policies. Here, the interdisciplinary cooperation or at least the well-meant work separation among policy-relevant disciplines such as social sciences, natural sciences and engineering are of high relevance. However, with the concrete analysis of the problem in the background and possibly effective solution-strategies, the contribution of a poli-

---

much to explanatory and predictive theory, neither have the natural sciences contributed to the reflexive analysis and discussion of values and interests" (Flyvbjerg 2001: p. 3).

61 See for example Agenda 21, Chapter 1, Preamble, 1.1; 1.3; Chapter 2, 2.1; 2.2; Chapter 34, 34.1; 34.2 etc. (http://www.igc.apc.org/habitat/agenda21 – 06 June 2010).

62 See World Summit for Sustainable Development (WSSD): Plan for Implementation – (http://www.johannesburgsummit.org/html/documents/summit_docs/2309_planfinal.htm – 17 July 2010).

cy-analysis through political science can be to identify and analyze the factors leading to answers, why a political system – here the case of Japan - does better or even worse on that score. It should thus be of interest how the recommendations are put into practice and why many of these are diluted or even never realized.

The general reason is that state programmes are usually not produced by one unitary actor holding all necessary resources for acting and whose interest is not exclusively the public welfare. Usually state programmes must be considered as the product of strategic interaction among several or even a large number of political actors, each having its own understanding of the nature of the problem and the realizability of certain solution-strategies (von Prittwitz 1994: p. 14). Additionally, each actor has its individual and institutional interests, normative preferences and resources for acting.

This study puts chief emphasis on the actors involved in transnational environmental policies. Although only individuals are capable of intentional acting, individual actors are typically acting with the interest and through the perspective of bigger units. Thus, for this study it appears justified to simplify the analysis through the limitation to some bigger units of complex actors. However, the analysis must still be able to revert back to the micro-level, if it appears empirically necessary.

Additionally, such bigger units as political parties, banks, NGOs, and ministries operate in institutional contexts, which restrict their action more than in the case of an autonomous individual. They are usually constituted of institutional norms defining their competences and resources, but simultaneously also prescribing the aims and thus the cognitive orientation. Restrictions of governance, but also chances for governance develop within such a field of tension (Braun 1993: p. 200). This means in the eyes of neo-functionalists that institutions, once they are created, show a high ability of insistence against changes.[63] Essential for this ability are the interests linked to the institutions. The more powerful such interests are and the stronger the institution's alliances with other actors are, the harder it is to replace them by another one (Colomy 1990: p. 483). The connection between complex systems and stable actors' interests leads to relatively closed social areas, which are not accessible without any efforts for governance. The interest to save the gained chances of power and the chances to effectively influence information impede governance. But the continuous battle for power of these actors also gives e.g. governments the chance to be successful with their interests, through coalition building and sanctions. In this case policy makes use of the openness of these social processes for power (Schimank 1992:

---

63 See for example Colomy 1990, 1985; Rüschemeyer 1974, 1977.

p. 172); actors also decide in accordance with their interests for a certain strategy of acting (Hohn & Schminank 1990: p. 29; Mayntz 1988: p. 31). The means by which institutes satisfy their interests is in most cases the possession of financial or other resources. Many actors, especially in development cooperation and environmental policy, are depending on outside financial resources like e.g. tax financing. The decreasing or increasing of the budget opens a certain chance of influence for political decision makers.

Nevertheless, political actors are usually in a relatively stable actor constellation. This applies especially to middle-grade governmental officials of e.g. ministries, who are not necessarily affected by the results of general elections and thus changes of governments, which are usually accompanied by the change of top-officials. To find answers to the key question of this study, it must also be analyzed which role the various actors have and whether the inter-actor constellations are responding to the new challenges. Furthermore, the institutional context defines the different types of interaction e.g. negotiations, hierarchical leading, unitary acting. Therefore, it must also be analyzed how different actors are influencing each other and the subsequent political consequences. This shifts the focus to the interdependencies of formal organizations and corporative actors and the character of the subsystem in which they are active (Mayntz 1988: p. 24). The character of this subsystem structures the access, processes, interests, framework conditions and influences of policies on the institutions.

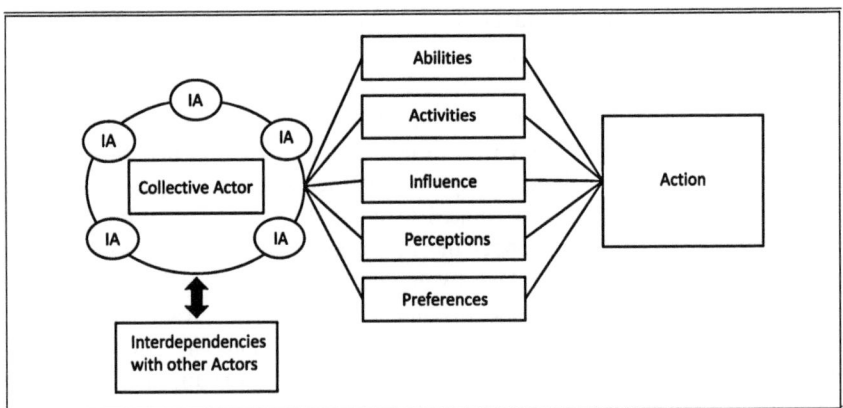

Figure 4: Simplified Model of Actors' Actions (IA = Individual Actor)
(Source: Own Illustration)

As this study aims to identify the set of interactions that directly or indirectly produces political results in the field of Japan and Germany's transnational environmental policies which are under analysis here, it tries to separate these interactions from each other, although they finally influence the result. This set finally makes up the research unit. Then in the second step, the aim is to identify different actors involved in the political process and having an impact on the final decisions. These actors are characterized by certain abilities, activities, influence, perceptions and preferences, which must be defined and analyzed; thus, it will be possible to work out the different resources for action of each actor. In the third step the actors' constellation and thus the interdependencies will be examined; here, it will be especially interesting in light of Dehnhard's neo-realistic theory to investigate if and how far national actors are influenced by international and non-governmental organizations. Can Dehnhard's theory prove that national actors even gain more power in the arena of environmental technology transfer? The relevance of the different actors cannot be worked out through their input on the policy-cycle. The interdependence of factual, organizational and legal areas is characteristic in establishing and carrying out a political programme. Policy analysis avoids focusing on any one of these areas, allowing the effectiveness of a particular organization in solving the problem to be examined, rather than treating the programme creation and realization as an isolated problem.

Consequently, a combination of the actors' concepts with the policy-cycle model appears useful for the methodological approach of this study.

7.2 Policy-Cycle

The policy analysis concentrating on the output of political processes is a relative newcomer (Böhret 1988: p. 7f) in the realm of political science, which has traditionally concentrated on the fields of polity (the study of institutions and democracy) and politics (political processes and behaviour). With the launch of the journal *Leviathan* in 1987, Japanese mainstream political studies of the time were criticized, arguing that Japanese politics was studied only marginally by researchers in the field of history, foreign studies and history of thought, based on impressionistic, subjective attitudes (Taniguchi 2008: p. 3f). Today major interests of Japanese political scientists are with history of European political thought, history of Japanese politics, local governments and administration and public administration and comparative politics studies, whereas policy analysis is not yet in the focus. In contrast to Japan after lengthy debates[64] in the 1980s as to its scientific standing, policy research has established itself in the German-speaking research world as the main analytical method (Schubert 1991: p. 187).

---

64 For a more detailed view of this discussion compare with Héritier (1993): p. 9-36.

The chief aim of a policy analysis is best summarized by Thomas R. *Dye*:

"Policy-analysis is finding out what governments do, why they do it, and what difference it makes." (Dye 1976: p. 3).

The main focus of policy analyses has been on topics of regional policies and on policies relating to technology, energy, employment, economy and the environment (Löbler 1990: p. 7). This focus on the problems typical for wealthy nations appears problematic because it takes a functioning state machinery for granted as well as the underlying possibility that the existing problems can be solved, in particular economic difficulties. However, as already indicated above, states are confronted with new challenges by e.g. transnationalization and this led to a debate on the states' ability to effectively respond to the new circumstances. Consequently, such an approach demonstrates in the words of Volker *von Prittwitz* "*a repression of reality, self-disillusionment, and at least in the medium term, a loss of credibility*" (Prittwitz 1993: p. 351) in political analysis.

Prittwitz's critics mirror the call of many scientists, including Georg *Jellinek* and Albrecht *Dehnhard*, to widen the perspective of analysis through a more multidisciplinary and holistic approach. Irrespective of social, political or economic structures, all human-beings require a continuous and undiminished supply of untainted natural biological materials to satisfy their food, shelter, energy and medicinal needs. Any future social systems will therefore have to closely heed the limiting biological factors, while establishing equitable and viable patterns of life throughout the world. Consequently, all analysis under the paradigm of sustainable development must also take the demanded multidimensional, holistic approach.

This study acknowledges Japan as one of the most advanced countries in the world, but will attempt to examine, through a policy analysis, the concrete problem of environmental technology transfer as part of its transnational environmental policy. It will not only try to meet the challenges through a multidimensional approach and to direct the focus to an analysis of the respective transnational environmental and development policies but the main emphasis of this analysis is on policies. However, as in political reality the three aspects of form, process and content work together; a policy analysis puts them in relation to each other.

Transnational environmental policies and thus also e.g. eco-political Official Development Assistance (ODA) can only be effective when it goes beyond the realm of explanations and programmes and is realized as intended by the Plan for Implementation of the WSSD and UNEP's Bali Strategic Plan for Technology Support and Capacity Building. During this process various realization elements or phases can be observed in which the various actors play different roles.

If these phases are pictured as forming part of a close series, the result is the policy-cycle (Prittwitz 1990: p. 93), a model of an iterative process.

That in turn allows policies to be viewed as a process of problem solving, which can be divided into different sequences (Mayntz 1982: p. 74ff). Beyond the multiplicity of terminology referring to individual stages of development of a policy, a consensus emerges to distinguish between the following phases:[65]

The first phase deals with the initiation and estimation. The initiation involves the identification of a political problem, following appropriate political examples, so that the problem becomes relevant in terms of action. In the second step of the first phase, the problem is analyzed, so as to set goals and define aims by making concrete suggestions for action.

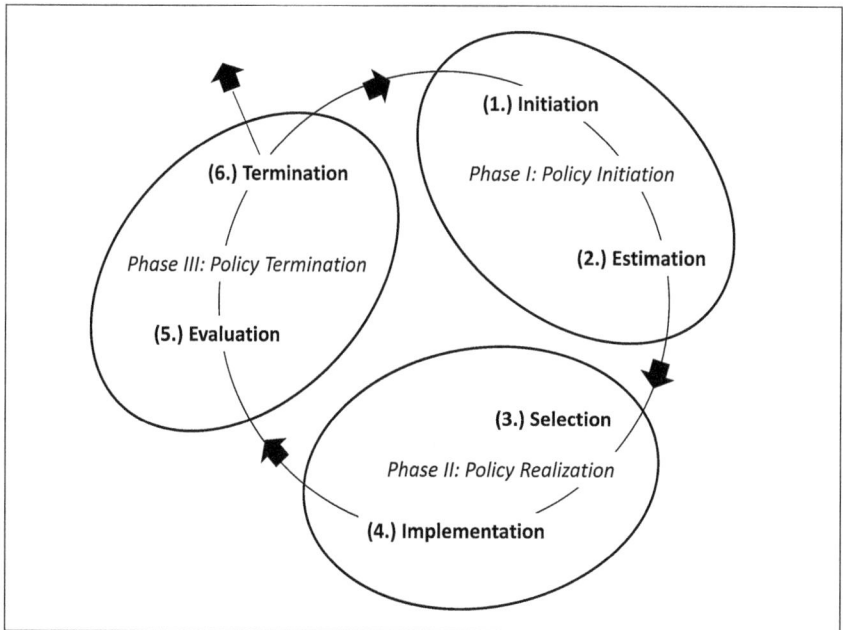

Figure 5: Dynamic, open Policy-Cycle Model
(Source: Own Illustration)

---

65 The policy-cyclus model was created on the basis of the Input-Output Models of David Easton. Accordingly, the political, administrative system converts social demands (inputs) into political acting (output) (Easton 1965).

The selection, step number three, involves the final decision on possible solutions to the problem debated in the previous steps. It is followed by the fourth step, the so-called implementation, in which political intentions become hard reality. The selection and implementation together form the second phase of the policy-cycle- the political realization.

The evaluation and termination both belong to the third phase. The termination, step six, where decisions are met about how the programme can be maintained, and whether or in what way it could be modified, evolves out of the fifth step, the evaluation. Actual termination seldom occurs, as in political reality, the chief decisions are concerned with maintaining existing programmes and how to modify them if necessary (Brewer 1983).

This division of the policy process into various phases provides the framework for many investigations. However, the phases of the problem-solving process do not necessarily follow each other as the model suggests. In the evaluation of alternative plans and the concrete formation of programmes for instance, questions regarding the implementation must be taken into account. Due to this interdependence of the separate sequences both among themselves as well as between concurrent political activities (policies), the closed circle of the Policy-Cycle Model must be depicted as a dynamic, open model (Rittberger 1985: p. 208).

Japan and Germany are among those more than 178 governments who adopted the Agenda 21, the Rio Declaration on Environment and Development, and the Statement of principles for the Sustainable Management of Forests at the United Nations Conference on Environment and Development (UNCED) held in Rio de Janerio, Brazil in June 1992, which are highlighting the need for environmental technology transfer as one key-aspects to fight human impacts on the environment. The Plan for Implementation of the WSSD suggests somewhat more concrete action on environmental technology transfer in its paragraphs 4, 14f, 19a & l, 45sexties and 100c. Nevertheless, it must be scrutinized how these comprehensive plans are practically implemented – globally, regionally, nationally and locally – by the respective governments.

With direct reference to the theoretical and the thus-far developed methodological approach of this study, a policy analysis with the help of the policy-cycle model also reflecting Flyvbjerg's four value-rational questions as part of phronetic social sciences (Flyvbjerg 2001: p. 65) appears suitable to find out:

- Which actors do what, when, why and how,
- Which part these actors play in the policy process,
- Which clues can be found within this analysis to verify or disprove Dehnhard's theory forming the basis of this work?

7.3 Comparison

To determine what works and what does not, as well as to demonstrate important theoretical relationships among variables, one has to make use or even develop the methodologies for entering into the real world of policies in a number of different settings and extracting meaningful information. Consequently, the information must not only be descriptively accurate, but those statements must also link observations made in one policy-cycle with similar observations made in another, or link the observations in one system with general propositions about policies (Peter 1998: p. 3). One country may be the crucial case for testing a hypothesis – if the proposition works there then it might work anywhere – so that the case study is one efficient mean of addressing the theoretical proposition. But in a more comparative mode of analysis, as is intended here, the similarities as well as the differences become evident, and the researcher must think more clearly about the root causes of the performances of the system.

A long and honourable strand of comparative political analyses among various other types[66] is to select a small number of instances of a process or institution that appear similar, or at least appear comparable in some important ways (Hague 1998; Berg-Schlosser & Müller-Rommel 1997; Hartmann 1995). This study focuses on Japan, but also provides certain comparison with Germany, and especially those actors involved in transnational environmental policies especially through the transfer of environmental technology to NICs. This study will also develop a certain comparison of Japan and Germany's particular institutions involved in environmental technology transfer in the 1990s. Furthermore, the comparative analysis of transnational policy formulation and implementation is a clear case of applying a process model, beginning with agenda-setting and ending with evaluation and feedback – thus the policy-cycle model. In consequence the intended comparison can be made on how this process unfolds in the different settings in Japan and Germany. This comparative element in this study proves or disproves Dehnhard's neo-realistic theoretical argument that states are gaining more power under the present challenges. Additionally, the policy-cycle model can illuminate the nature of the policymaking process and the actors themselves enabling one to learn about the impact of new challenges on the state's policies.

---

66 E.g. (1) single descriptions (Ramage 1995, Rose 1989); (2) typology studies for countries or subnational units (Bebler 1990); statistical analyses from a subset of world's countries (Kaase and Newton 1995) and statistical analyses of all countries (Sullivan 1996).

Within the methodic framework of a comparative, open and dynamic policy-cycle-model and likewise a comparative actors' analysis based on phronetic social sciences, the question is how this concept can be used in a policy analysis?

7.4 Empirical Research

Empirical social research requires the survey or production of relevant data, the processing and analysis of those to receive information, their storage as basis for comparison and learning, and the presentation of the gained knowledge e.g. for scientific communication. Thought patterns and the patterns of perception of the analyst as well as the definition of the key questions, the research spectrum and the structural concepts influence the categorization of information or knowledge respectively.

In a policy analysis the research process is often understood to involve a subject-object relationship between the researcher and the findings (Kritz 1990: p. 84). But within social scientific research, human-beings and their products are the research objects and thus can directly or indirectly manipulate the research work. For example as an actor or expert they are often integrated into the collection of relevant data through e.g. interviews, in which their responses are also according to their own values.

This leads to an even increased complexity of the research, since on top of rather simple subject-object relationships one has to include interdependent subject-subject relationships between the researcher and the social actors (Heinze 1995; Spöhring 1989: p. 11). One can suppose that e.g. interviews of a German conducted in Japan and thus in a different culture bring along not only linguistic obstacles, but also intercultural problems at the expenses of important information. Both a detailed and comprehensive preparation, transparency in the research work and certain confidence-building measures can help to minimize an intervention in the analysis.

The three fundamental questions mentioned by Dye in his definition of policy analysis, "what, how and why", reflecting the four value-rational questions of Flyvbjerg provide a guide as to three possible approaches to solving the problems. The "what" question forms the statement of empirical analysis (Schubert 1991: p. 43). The "how" question will be analyzed using the above-developed applied combination of methodologies, and the creation of a chain of cause and effect for the problem under investigation. The "why" question, provides an explanation on the connection between the answers to the "what" and "how" questions. The ex-post perspective examines retrospectively the outcomes of state politics and related issues (Windhoff-Héritier 1987: p. 116). Finally, all of this should also be done by addressing the value-rational questions at its core and by ensuring that the public has use for the answers in their deliberations about praxis (Flyvbjerg 2001: p. 163).Against this background this analysis must be defined as an empirical study (Gerlach 1988: p. 154).

Additionally, the above-described limitation of studies on transnational environmental policy studies through the transfer of environmental technology with special focus on the donor countries led to a complex reference situation. Based on the methodological approach for this study, various sources served as basis for this research work:

- Relevant natural-scientific publications on environmental problems,
- Official reports of the Japanese and German Government and Parliament (or such which were conducted on behalf of those through enquete commissions or consultancies),
- Publications of international organizations (e.g. UNEP, UNDP, UNIDO, UNU, OECD, World Bank; Asian Development Bank etc.), research institutes and NGOs (e.g. Friends of the Earth),
- Press reports,
- Relevant social science secondary literature,
- Archives,
- Author's discussions with experts.

These discussions are a classical tool of empirical research and were essential for this study, also being grounded on phronetic social sciences. The large majority of the interviews took place between May 1995 and June 2000 although some interview partners were re-contacted in 2009 and 2010 for the ex-post review of their policy assessments in the late 1990s. The selection of experts (N=42) was made at the beginning of this research work on the basis of the available literature and enlarged step by step through the snowball-effect. The interviews lasted 30 to 240 minutes or were performed through extensive exchange of emails. The author met some of the experts several times to develop a certain mutual confidence, and in some cases they were at the author's disposal for further inquiries. The group of experts is made up of social and political scientists, environmental and natural scientists, politicians, consultants, present and former governmental officials, industry representatives, NGOs and staff of international organizations. The discussions or interviews were realized with a non-structured and only partly-standardized questionnaire to be able to listen to the interviewees carefully and to obtain the maximum amount of information through discussion. Most of the meetings were recorded when it was acceptable for the interviewee; otherwise notes were made; the emails were archived for reference.

7.5 Qualitative and Quantitative Methods

The above described empirical approach brings along two key questions of social sciences about the principal possibilities to (i) value-free research and (ii) an objective picture of the social reality within the framework of empirical research. They are among other things reflecting the search of the social sciences to develop their own scientific identity to set off against natural sciences with their exact models of measurement. Additionally, this mirrors the call for the application of social scientific methods of understanding to what extent the acting hits the target (qualitative dimension) and to the identification with or the transfer of quantitative, explaining approaches respectively.

Methods used for empirical research as in this study, integrate quantitative steps in the research process, for example to illustrate e.g. the necessity, the success and monetary value of the above-mentioned measures through the introduction of appropriate technologies to reduce the emission of CO_2 (Naßmacher 1991: p. 182). Quantitative and qualitative methods are of high relevance for the analysis of environmental policies along the paradigm of sustainable development. The final qualitative analysis of Japan's transnational environmental policies through the transfer of environmental technology to NICs, likewise, requires a quantitative analysis.

Part B – Background

As illustrated above the transfer of environmental technology into NICs as part of Japan's transnational environmental policy proceeds on the assumption of the existent, but probably reduced capacity of the nation-state to act. Therefore, the following will pay special attention to the actors and how they were influencing Japan's transnational environmental policies in the 1990s. It will be important to find out how far they were influenced from other non-national actors and which were relevant factors for the policy formulation and implementation process.

8 History of Transnational Environmental Cooperation

The number and range of international agreements on environmental practices and policies have grown in recent years. Between 1990 and 2002 the number of multilateral and international environmental conventions increased by at least thirty following the chronological index of the Fletcher School.[67] According to an estimate by Peter M. *Haas,* Robert O. *Keohane* and Marc A. *Levy* (1993), of 140 international environmental agreements instituted since 1921, more than half entered into force after 1972. Edith *Brown Weiss,* Daniel B. *Magraw* and Paul C. *Szasz* (1992) estimated the total at a much higher than 900 agreements in force since 1992, including regional and bilateral treaties.

Transnational environmental cooperation did not become a part of international relations until the Stockholm Conference on Human Environment in 1972 with one of its outcomes being the foundation of the United Nations Environment Programme (UNEP). Transnational environmental problems have been on the political agenda since the beginning of the 20th century and thus motivated national governments to find agreement in coping with the respective challenges.

Today, multilateral and bilateral environmental treaties are dealing with nearly every kind of transnational environmental problems. Not only the existence of such contracts makes clear the changed consciousness of justice and norms, but also the perception of environmental problems and political strategies for their solution. Nevertheless, the environment did not necessarily benefit from all of them (Simonis 2002b: p. 74; 1999). Remarkable is the predominance of marine, lake and river-related agreements. In the truest sense of the word, border lines are fluid in water and thus permeable for every kind of pollution.

---

67 See http://fletcher.tufts.edu/multi/chrono.html – 10 June 2010.

Moreover, seas, lakes and rivers always represent a sensitive, indispensable, but limited source of food for a large part of the world population. In this sense (over)fishing in international waters required transnational conventions, ultimately leading to common efforts for the protection of the water and its flora and fauna.

For the analysis of Japan's transnational environmental policies, it is of interest to investigate the development of transnational environmental cooperation within this chapter and thus to find out more about e.g. actors, priority fields, motives and experiences that might be of relevance for the present formulation and implementation of policies.

8.1 Multilateral

The Agreement for the Protection of Birds Useful for Agriculture[68] from March 1902 can be considered as the first transnational environmental agreement between 16 European states including Germany. The purpose of this agreement was to extensively support the well-being of birds through e.g. the dam up of the mass murder of migrating birds, the creation of nesting places and feeding in winter. Impetus came among others from the German Society for Bird Protection (*Bund für Vogelschutz (BfV)*) with Lina *Hähnle* as its chair. Mrs. Hähnle, wife of the entrepreneur and liberal member of the German *Reichstag* Hans *Hähnle* aimed at a broad natural conservation under the flag of bird protection (NABU 1999: p. 2).[69]

Germany was also a party of the Police Regulation of Fishing in the North Sea out of the Costal Waters (1882) and the Contract Concerning the Regulation of Salmon Fishing in the Rhine River (1886). Although these agreements deal with limited natural resources, the economic interests clearly prevailed ecological consideration. The securing of certain national fishing quota triggered these transnational contracts, instead of joint efforts against over-fishing and thus a massive decline of the fish stocks, which is still a major challenge in transnational environmental cooperation these days, as the fish stocks are dramatically shrinking and the share of young fish in the nets unable to reproduce, is exten-

---

68 Translated by the author. The German title of this agreement is Übereinkunft vom 19.3.1902 zum Schutz der für die Landwirtschaft nützlichen Vögel.
69 The foundation of the German Society for Bird Protection (Bund für Vogelschutz (BfV)) was basically a reaction to the industrialization with its enormous impacts on the environment and the social structure. The BfV was a Wilhelmina reform movement, which sets a backward oriented worldview, idealizing the agrarian society against the modernity.

sively increasing. And even though UNEP's chronological summary file shows the Convention Concerning the Use of White Lead in Painting (1921) as the first environmental treaty, the objective of this agreement to protect workers from the exposure to white lead and lead sulphate and all products containing these pigments falls into the category of health; however, neither Japan nor Germany participated in this treaty.[70]

In November 1933 it came to the first classical environmental convention – the Convention Relative to the Preservation of Fauna & Flora In Their Natural State. This convention was primarily decided to introduce a special regime for the preservation of flora and fauna, particularly of Africa, through e.g. the constitution of national parks. Again, neither Germany, whose colonies were distributed among the victorious nations of the First World War, nor Japan, which was never in command of any territories in Africa, were among the signing governments. Also the Convention on Nature Protection and Wild Life Preservation in the Western Hemisphere (1940) is outside Japan's and Germany's interests, as it is an agreement of American republics.

To maintain and increase international cooperation in controlling pests and diseases of plants and plant products and in preventing their introduction and spread across national boundaries, Japan and Germany ratified the International Plant Protection Convention (1952); Among others Japan and Germany also became parties of the Convention on the High Seas (1958), The Antarctic Treaty (1959), Convention on Wetlands of International Importance Especially as Waterfowl Habitat (Ramsar Convention) (1971), and the Convention for the Protection of the World Cultural and Natural Heritage (1972).

It was only with the Convention on Long-Range Transboundary Air Pollution in 1979 that the first multilateral environmental treaty, which devotes transnational environmental problem of air pollution in Europe, was signed. It was chiefly the growing acid deposition in Europe and North America with far-reaching impacts on drinking water, fish and other aquatic life forms, with threatening impact on forests that led to this treaty. The convention expressed a common commitment on the part of the signing parties to reduce acid precursor emissions and jointly monitor emissions and precipitation chemistry. Yet it is amazing that with a history of more than 30 years of experience in developing regimes to handle transnational environmental problems and with cases in industrial countries of partly catastrophic air pollution not necessarily halting at national border lines,[71] clearly demonstrating a connection between extremely high

---

70 http://sedac.ciesin.org/entri/summaries-menu.html – 10 June 2010.
71 In winter 1952 about 4,000 people died because of smog resulting from the burning of low-cost but high sulphur coal in nearly each household in London.

levels of pollution and day-to-day excessive mortality (EPA 1996), the first multilateral environmental agreement on air pollution was signed only at the end of the 1970s.[72] Until then the fundamental conviction of the majority of decision makers from politics and industry was in line with that of the people, namely that national initiatives are only necessary to prevent extreme situations through air pollution. However, smoking chimneys were still regarded as symbols of growth and prosperity, so that drastic countermeasures put these at risk. Furthermore, national initiatives were for a long time regarded as sufficient and easier to channel and control; this assumption turned out to be a fallacy as experiences in the mid-1970s with the dying of trees which was one of the societal drivers moving the issue into the public focus. Ultimately, there was a three part reportage in the German weekly magazine *Der Spiegel* (November 1981) under the title: "Acid Rain: There is something in the air"[73]. Conspicuously, the acid rain caused by air pollution in some affected regions had to have originated from across the border, because of the wind directions. Japan also experienced effects from acid rain in the mid-1990s, resulting from e.g. high sulphur lignite burning in China, with effects along Japan's Western coast.

Further international agreements signed by Japan and Germany are the International Tropical Timber Agreement (1983), the Protocol on Substances that Deplete the Ozone Layer (1987), the United Nations Framework Convention on Climate Change (1992), UNEP's Basel Convention on the Control of Transboundary Movements of Hazardous Wastes and Their Disposal (1989) and the various agreements forged as part of the United Nations Conference on Environment and Development (UNCED) in 1992.

On the multilateral level, two cases will be analyzed in more detail in the following, as either Japan and Germany or one of both is playing a major role here (i) the Whaling Convention and (ii) the Kyoto Protocol.

Additionally, water pollution through e.g. ammoniak, chlorine, salpeter, was recognized as a major environmental burden since the beginning of the century and resulted in many diseases including a typhus epidemic in 1901 in Gelsenkirchen. In the winter of 1962, the air pollution was worsening the situation in Germany's Ruhr Valley with a sight of less than ten metres during some nights resulting in increasing mortality rates.

72   In 1285 London's air was so polluted by coal burning, that King Edward I established the world's first air pollution commission. Twenty-two years later the same King made it illegal to burn lignite; one Londoner was even executed for doing so. But the ban as such was not successful.

73   German original title: Säureregen: Da liegt was in der Luft.

8.1.1 Case of the "Whaling Convention"

The International Convention for the Regulation of Whaling (1946) aiming to protect all species of whales from overfishing and safeguard for future generations as well as to establish a proper system of conservation and development of whale stocks was the first multilateral environmental agreement that entered into force in Japan in 1951. Germany, together with many more primarily non-whaling nations, was in agreement with the outcomes of this convention 31years later in 1982, when the International Whaling Commission (IWC), despite no agreement from its Scientific Committee, agreed on a moratorium on commercial fishing of all whales from 1985.[74,75] Japan, Norway, Iceland and few other nations have voiced objections against this moratorium but the US placed pressure on especially Japan by using the Packwood-Magnuson Amendment. This Amendment to the US Fishery Conservation and Management Act of 1976 allows the United States to reduce or suspend fishing privileges in US waters for nations acting contrary to IWC guidelines. As Japan was concerned about its own fishing industry and its enormous trade surplus with the US at that time (JWA: 1988), Japan withdrew the objections from the IWC Convention and terminated commercial whaling. In spite of the US promise to refrain from imposing sanctions on Japan, the US executed the Packwood-Magnuson Amendment on Japan in 1988. This political decision reflected a cultural bias that whales are not resources which should be exploited; although Japan officially accepted the ban of commercial whaling, it has been accused numerously of violating the moratorium.

The International Convention for the Regulation of Whaling contains a provision allowing member nations to self-issue permits to kill whales for scientific research. When the commercial whaling moratorium took effect, Japan, Iceland and Norway issued themselves research permits. In the case of Japan, Japan conducted its Whale Research Programme under Special Permit in the Antarctic (JARPA), killing about 100 ± 10% minke whales annually between 1994 and 1999 (Government of Japan: 2002). Starting at the beginning of 2006, JARPA II was launched as a comprehensive long-term research programme where an interim detailed review is to be conducted following completion of the first six

---

74 The Scientific Committee asserted the moratorium as unnecessary (Nagasaki 1995).
75 The moratorium was decided on the condition that in 1990 at the latest a comprehensive check should be conducted to analyze the effects. On the basis of this a modification of the moratorium or new fishing quota should be discussed. But the IWC decided against these discussions and confronted whaling nations with a new ban of commercial whaling. Thus, the Norway government, an active member of the IWC since 1946, resumed whaling again in 1993.

years of research. JARPA II research will focus on Antarctic minke whales and two larger whale species, namely humpback and fin whales. The number of Antarctic minke whales to be sampled is 850 ± 10% during JARPA II. In addition 50 fin and 50 humpback whales will be sampled.[76] Japan argues that the whale species to be sampled in JARPA II have recovered to a robust or nearly robust stock, so that the research poses no risk to them. Moreover, the research is, in Japan's eyes, necessary to understand where the changes in the Antarctic marine ecosystem are leading and will be demonstrated by JARPA. This should help to study appropriate management methods and ways of utilization in the future.[77]

However critics insist Japan's research programme must be seen in a larger context and that it is merely an approach to secure unlimited access to global marine resources. For a long time supplying restaurants with whale meat was prominently referred to (Greenpeace: 2002); in the US and Europe, consumption of whale meat is discouraged due to the moral and philosophical outlook that developed over time. Popular movies like *Flipper* and *Free Willy* also contributed to the presentation of whales as the most specialized mammals, being sentient, intelligent, having their own community and effectively demonstrating that whales also experience emotion such as suffering (Skare: 1994).

The Japanese always have taken whales for food due to the Buddhist belief that people must not eat meat from "four-legged animals" until the middle of the 19th century. Additionally, the Japanese culture developed in close relation with the sea and its resources. However, since whales were regarded as fish by the Japanese, they also caught whales for the consumption of their meat; therefore, whaling is considered as an integral part of Japan's history and culture. Unlike Western countries where whales have been caught and killed for products like oil, Japanese whaling has been predominately for human consumption (Japan Whaling Association: 2002). Accordingly, the Japanese have different views on whale conservation, viewing whale exploitation not much differently from cattle

---

76 According to the Whale and Dolphin Conservation Society (WDCS), IWC and the International Union for Conservation of Nature (IUCN), Japan killed 160 Pacific minke whales and 506 Antarctic minke whales in 2009 alone. In addition Japan took 100 sei whales, 50 bryde whales and each one pott and one fin whale in the same year (Sueddeutsche Zeitung: 2010).

77 For more details see e.g. the Japanese Government's position on whaling (http://www.au.emb-japan.go.jp/pdf/Whaling.pdf – 02 July 2010), the Japanese Institute of Cetacean Research (http://www.icrwhale.org/eng-index.htm – 02 July 2010), the Japan Ministry of Foreign Affairs (http://www.mofa.go.jp/policy/economy/fishery/index.html – 02 July 2010) and the Japanese Fisheries Agency (http://www.jfa.maff.go.jp/e/whale/index.html – 02 July 2010).

consumed by US citizens or Europeans, though nowadays whale meat is only consumed by a very small minority of Japanese. Japanese officials are increasingly campaigning to resume whaling, especially because of estimates on minke whale population abundance (760,000 in the Antarctic and 118,000 in the northeast Atlantic, respectively) from Japan and Norway. This indicated in the eyes of the officials that these populations could sustain a limited commercial harvest (Buck: 1997); yet the Japanese estimates are in question. The IWC explains that it is unable to provide reliable estimates and that its Scientific Committee is undertaking a major review (IWC: 2010).

At the meeting of the International Whaling Commission (IWC) in June 2006 in St. Kitts and Nevis, a declaration[78] calling for the eventual return of commercial whaling was passed by a majority of just one vote; since then Japan has threatened to pull out of the IWC. Just before the IWC annual meeting in 2010 where a compromise was expected between pro and anti-whaling countries, including commercial whaling on a limited scale, Fisheries Minister Yamada threatened once again. The proposal to allow commercial whaling has drawn criticism from all sides bringing fresh attention to the whaling issues (Alabaster 2010). However, it must be noted that Japan and other pro-whaling nations did not succeed in ending the moratorium during the 62nd IWC Meeting in June 2010 in Agadir, Morocco. The IWC discussed a proposal including the retaining of the 1986 moratorium on commercial whaling and establishing catch limits for whales that will reduce catches over 10 years. Additionally, indigenous subsistence whaling will still be allowed. This proposal would have allowed Japan to resume whaling in its coastal waters but reduce its catching of whales in the Southern Ocean. Not even the strong push by a few of the conservation-minded nations and several NGOs could neutralize the scepticism and ultimately, the opposition of other parties. In the end no one could really determine whether or not Japan was ever willing to find a compromise (Unti: 2010).

Moreover, corruption allegations overshadow the annual meeting: The IWC's deputy chair Anthony Liverpool (Antigua & Barbuda) accepted free flights and had his hotel paid by the Japanese government, claimed Britain's *Sunday Times* newspaper. This would be a break of International Convention for the Regulation of Whaling[79]

NGOs including Greenpeace, WWF and Pro Wildlife accuse Japanese officials of bribery especially with reference to developing countries through promises of Official Development Assistance (ODA), but especially fisheries aid, and thus to support Japanese efforts to abolish the moratorium (Sueddeutsche

---

78 See http://www.whaling.jp/english/history.html – 28 March 2010.
79 See http://www.iwcoffice.org – 01 July 2010.

Zeitung 2002; Tageszeitung 2002; n-tv 2002: 21 May 2002; Der Spiegel 2010: 21 June 2010; Die Zeit 2010: 23 June 2010). There is no evidence supporting this linkage between votes for Japan and the aid money provided, but a number of statements from those involved or closely observing the IWC negotiations and Greenpeace's briefing on financial grants and votes for whaling (Greenpeace: 2007).

Despite all ideological discussions, the threat of some species of whales is without quest and accepted by all whaling and anti-whaling nations. It is also without quest that the limited harvests of Japan, Norway and Iceland these days are not seriously endangering the whale populations, but the growing pollution of the oceans.[80] However, the slow recovery of the whale population under harder environmental conditions would come to an abrupt end with an official legalization of commercial harvests. A great demand would also lead to a growing number of poachers who would ignore all quotas and are difficult to apprehend in violating international agreements.

These problems also illustrate the increasing pressure from actors outside the national territory influencing or at least trying to influence national policies in certain directions. This way of transnationalization results in a countermovement that highlights the endangerment of cultural and national characteristics. They become a pawn in a game dominated by a hegemon and thus consequently lose the ability to act for and efficiently control the well-being of a nation.

Obviously, the whaling dispute could be one argument against Dehnhard's theory, which is under examination here, because Japanese politicians are confronted with immense pressure from actors outside their country. On the other hand this pressure has led to domestic and international discussions and fostered a certain sensitivity regarding the loss of national and cultural characteristics but also unlimited access to limited resources such as whales indicating that the Japanese may not necessarily be re-gaining power in other political fields and spheres again.

8.1.2 Case of the "Kyoto Protocol"

As an approach to stabilize atmospheric concentrations of greenhouse gases at safe levels, the United Nations Framework Convention on Climate Change (UNFCCC) came into force in March 1994. Parties of the UNFCCC meet regularly at the annual Conference of the Parties (COPs) to review the implementation of the convention and continue talks on how to best tackle climate change. Decisions made during the COPs make up a comprehensive "rule book" for the

---

80 In 2009 Norway killed 81 minke and 125 fin whales; Iceland 484 minke whales (Sueddeutsche Zeitung 2010).

8 History of Transnational Environmental Cooperation

implementation of the convention and key-decisions are sometimes given a more high-profile title - such as the "Berlin Mandate" and the "Kyoto Protocol".[81]

The Kyoto Protocol was adopted during the COP3 in the former Japanese capital of Kyoto in 1997 and entered into force in 2005. The detailed rules for the implementation of the Protocol were adopted at COP7 in Marrakesh in 2001, and are called the Marrakesh Accords. The Kyoto Protocol aims to limit or reduce the emissions of the six greenhouse gases namely carbon dioxide (CO_2), nitrous oxide (N_2O), methane (CH_4), hydrofluorocarbons (HFCs), perfluorocarbons (PFCs) and sulphur hexafluoride (SF_6) to individual, legally-binding targets adding up to a total cut of at least 5% from 1990 levels in the commitment period 2008-2012.[82] Some specified activities in land-use, land-use change and the forestry sector, namely afforestation, deforestation and reforestation, which emit or remove and thus absorb CO_2 from the atmosphere, are also covered. Additionally, the protocol establishes three innovative mechanisms: (i) joint implementation, (ii) emissions trading and the (iii) clean development mechanism, which are designed to help parties to reduce the costs of meeting their emissions targets by achieving or acquiring reductions more cheaply in other countries than at home. The clean development mechanism aims to assist developing countries in achieving sustainable development by promoting environmentally-sound investment in their economies from industrialized country governments and businesses.

Japan as president of the COP3 failed to guide the negotiations to success in order to prevent global warming. In August 1997 four months before the COP3, the German Minister for Environment, Angela *Merkel*, and the Japanese Direc-

---

81 Parties agreed that specific commitments in the convention for Annex I parties were not adequate. They launched a new round of talks to decide stronger and more detailed commitments for these countries.

82 Table:

Country	Target
EU-15, Bulgaria, Czech Republic, Estonia, Latvia, Liechtenstein, Lithuania, Monaco, Romania, Slovak Republic, Slovenia, Switzerland	-8%
USA	-7%
Canada, Hungary, Japan, Poland	- 6%
Croatia	- 5%
New Zealand, Russian Federation, Ukraine	0
Norway	+ 1%
Australia	+ 8%
Iceland	+10%

(Source: Kyoto Protocol of the UNFCCC, Annex B)

tor-General of the Environment Agency, Michiko *Ishii*, participated in a Japanese-German Environmental Symposium in Tokyo. During her speech Mrs. Merkel asked Mrs. Ishii to clarify Japan's positions and provide a clear proposal as the president of the COP3, and not to wait for the USA's opinion and to closely cooperate with Germany. Thus, Mrs. Merkel was referring to Japan's choice to develop individual measures while considering the reaction of USA to each. However, Mrs. Ishii was only in the audience, listening to the statements of panelists, not taking the chance to unveil Japan's own position. In preparing the COP3 the Japanese government emphasized a figure of 2.5% reduction in CO_2 emissions, because it concluded that the US government would not accept a reduction of 5% in CO_2 emissions due to its resistance against any increase in gasoline tax. But Europe and the Alliance of Small Islands States still insisted on large-scale reductions compared to figures of 1990, not even taking Japan's proposal into consideration.

During the COP3 the new Director-General of the Environmental Agency, Hiroshi *Oki*, aimed at resigning as the president and returning to Tokyo for lining up with the cabinet, which was confronted with a vote of no confidence in the Japanese Diet. Only some Diet members and NGO representatives urged him to change his mind and to return, but he was already far away from the conference site. Major decisions on finalizing the operational rules of the Kyoto Protocol were expected at the COP6 in The Hague, Netherlands. It was aimed to come to final decisions on how much of a nation's commitment could be met through emissions trading systems, consequences for non-compliances, and to which extent one could take credit for carbon absorbed in the sinks. Unfortunately, the COP6 negotiations collapsed when the United States and the European Union reportedly led by Denmark and Germany failed to reach agreement on these key-issues, with particular controversy centring on how much credit nations could be allowed for carbon uptake – sequestration – by forests. In late March 2001 the Administration under US President George W. *Bush* considered the Kyoto protocol "dead" in terms of US policy. This initiated a high-level effort by Europeans nations with Germany at the forefront but also by Japan playing a major role to re-engage the USA in the Kyoto process, especially because the US emits a quarter of all CO_2 emissions worldwide or 40% of all industrialized countries.

To the surprise of many, negotiators resumed at the "COP6 bis" meeting in Bonn in July 2001 and during COP7 in Marrakech in November 2001 reached agreement on most of the outstanding political issues and thus provided a sufficient basis for the countries to ratify. Countries with commitments under the Kyoto Protocol to limit or reduce greenhouse gas emissions must meet their targets primarily through national measures. As an additional means of meeting these targets, the Kyoto Protocol introduced three mechanisms:

1. Emissions trading allow countries that have emission units to spare - emissions permitted them but not "used" - to sell this excess capacity to countries that are over the accepted targets. Thus, a new commodity was created in the form of emission reductions or removals. Since carbon dioxide is the principal greenhouse gas, people speak simply of trading in carbon. Carbon is now tracked and traded like any other commodity. This is known as the "carbon market".
2. Clean Development Mechanism (CDM) allows a country to implement an emission-reduction project in developing countries. Such projects can earn saleable certified emission reduction (CER) credits, each equivalent to one tonne of CO2, which can be counted towards meeting Kyoto targets.
3. Joint Implementation (JI) allows a country to earn emission reduction units (ERUs) from an emission-reduction or emission removal project, which can be counted towards meeting its Kyoto target. Joint implementation offers Parties a flexible and cost-efficient means of fulfilling a part of their Kyoto commitments, while the host Party benefits from foreign investment and technology transfer (UNFCCC 2010).

Both the Japanese and German governments represented their individual interests with stress and considerably influenced the negotiations and final decisions on key aspects of the Kyoto Protocol. Although both nations played a strong role in trying to re-engage the US in the Kyoto Protocol negotiations in order to ensure the effectiveness of measures against climate change, and although Japan and Germany agreed on not intending to delay the progress of the negotiations, Japan took an active part in softening the EU's stance on certain points. This request was obviously not only guided by the attempt to maximize the chances of bringing the US back to the Kyoto fold, but also to lighten the load through commitments out of the Kyoto Protocol for Japan. Therefore, Japan supported the US with the position that there should be no limits to the amount of emissions reductions that are allowed toward a country's obligations through emissions trading or joint implementation. Japan's Minister of Environment justified this standpoint with the argument that her country's CO_2 emission per Gross Domestic Product (GDP) is already among the lowest of the industrialized world (Kawaguchi 2001: 19 July). However, these efforts were to no avail and in mid-June 2001 President Bush outlined his preferred approach and confirmed the rejections of the US.

Japan's total consumption of energy per capita was less than 70% of the USA and less than 80% that of Germany in the late 1990s (OECD 1999: p. 221). This demonstrated in the eyes of the government that Japan has already achieved the highest level of energy efficiency; still this cannot obscure the fact that Japan's efforts to reduce greenhouse gas emissions have not gone well. Japan's greenhouse gas emissions fell in 1997 and 1998 due to the economic crisis and according to ministry officials, they exploded in 1999, with private households increasing emissions by 5.3%, from industry by 4.2%, from office buildings by 3.3% and by 23% in the transportation sector (Asahi News Service 2001: 11 July). This increase continued in the following years until 2006 and since 2007 again (OECD 2005: 54; IWR 2010). CO_2 emissions from energy use are by far the biggest source of greenhouse gases in Japan. Attempts to domestically intensify the measures, although it is officially intended by the government, cannot be positively assessed. This is due to its increases in CO_2 emissions in international comparison with e.g. Germany, France, and the United Kingdom which succeeded in reducing their CO_2 levels since 2001.

It was expected that Japan would extensively make use of the mechanisms under the Kyoto Protocol such as CDM and emissions trading against which EU Members and others offered long-time resistance in order to force nations to take more extensive domestic action to reduce emissions. There were three different indications for this:

1. The Japanese Minister of Environment never tried to emphasize the "Kyoto Initiative" as the COPs approached.[83] Under this initiative Japan has committed itself to providing on average annually US$ 2.4 billion in highly concessional loans for projects related to climate change in developing countries and furthermore, financial and technical assistance amounting to US$ 7.4 billion since 1998 (Kawaguchi 2001). Thus, Japan assigned its ODA through the provision of loans, the transfer of environmental technology and through environmental capacity building a major role in responding to the Kyoto process.
2. Throughout the negotiations Japan and Russia, assisted by Canada and Australia insisted on lax criteria. Russia, by far the largest credit supplier on the emissions market, was considered essential for filling Japan's gap in the Kyoto target through buying credits.
3. Japan has the obligation to reduce greenhouse gas emissions by 6% of 1990 levels for the first commitment period of 2008-2012. Observers considered this a rather low reduction target, which resulted mainly because

---

83 For a more detailed description on the background and philosophy behind the Kyoto Initiative please refer to MOFA 2000.

Japan's Ministry of International Trade and Industry (MITI) actively insisted that it is impossible to reduce CO_2 emissions below this, even if realistic policies and measures were implemented (Ayukawa & Mizutani 2002). Within the Japanese Government's New Climate Policy, the first period depended heavily on industry's current Voluntary Action Plan and excluded possibilities of any mandatory CO_2 emission reduction measures to be introduced before 2004. However, as Japan's CO_2 equivalent emissions did already increase by 6.8% by the end of 1999, this measure did not appear effective. So, this policy initiative was commonly seen primarily as a compliance with the industry. Additionally, this climate policy planned to expand nuclear power generation by 30% by 2010, which is associated with unresolved environmental problems and hampers the active development of renewable alternatives.

The developments since the COP7 proved that Japan is also extensively building on mechanisms such as CDM through the "Hatoyama Initiative" (see Chapter 1). In consequence Japanese businesses are intensively investing in greenhouse gas reduction projects abroad, mainly Asia. Firms that earn greenhouse gas emission credits through CDM projects abroad can count them in their reduction efforts and surplus credits can be sold through an emissions transaction on the market. This has opened another lucrative business. After lagging behind in its obligation to cut emissions, Japan is seeking emissions credits overseas e.g. through signing an agreement on buying surplus allowances for greenhouse gas emissions from the Czech Republic in September 2008 or starting such negotiations with Russia in March of the same year. The Basic Bill on Measures to Prevent Global Warming was adopted by the Japanese cabinet on 12 March 2010 and includes the goal to cut emissions by 25% by 2020 and 80% by 2050, compared to 1990 levels. It highlights emissions trading as a key policy to make these emission cuts reality; the bill is currently under discussion in the Diet. Yet after experimenting with Voluntary Emissions Trading for several years, it was only on 1 April 2010 that the Tokyo Metropolitan Government introduced the first real emissions trading scheme that will be implemented in Japan.[84] This indicates that a strong emissions trading scheme may have little impact on the Japanese economy, while allowing Japan to effectively reach its targets for emission reductions.

---

84 See The Guardian 2010: Tokyo kicks off carbon trading scheme Japanese metropolis launches Asia's first emissions cap-and-trade scheme, 8 April 2010 (http://www.guardian.co.uk/environment/2010/apr/08/tokyo-carbon-trading-scheme – 21 July 2010).

An international framework to succeed the Kyoto Protocol, which expires at the end of 2012, was expected to be adopted at COP15 in Copenhagen. Unfortunately, no decision was reached placing new focus of attention on whether the COP16 nations can reach a major political agreement in Cancun, Mexico that could lead to the adoption of a new protocol. During the COP15 the Japanese government showed a lack of leadership and visibility (Johnston 2009) as well as during the following talks in Bonn in May 2010 despite the ambitious "Hatayama Initiative".

In contrast to Japan Germany's role regarding the detailed operational rules for the Kyoto Protocol is less controversial. This results from Germany's stubborn position against the soaking of especially compliance measures and sinks rule which gave rise to criticism from e.g. the US and Japan referring to the German unification and the restructuring of the British coal industry as reasons for the success in reducing of greenhouse gas emissions in the EU. But a study of several German and British institutes for the German Federal Environment Agency (UBA) came to the conclusion that the greenhouse gas reductions in Germany and the United Kingdom are 50% a coincidence and 50% policy induced (UBA 2001). At the same time Germany and the United Kingdom were said to be on the right path to meet the Kyoto target as a part of Europe's climate change programme advocating a twin-track approach. This includes a multi-stakeholder consultative process for policy formulation and an internal EU greenhouse gas emission trading system. Furthermore, Germany has found its way in actively supporting renewable energy alternatives and additionally its way out of nuclear power generation within the next decades, though the present coalition of Christian Democrats and Liberals under German Chancellor Merkel have re-discussed this issue for a while now.

The results of the COP15 may have influenced the present government's decision to cut the support of the development of solar energy in case the nations would have agreed on the substantial reductions of CO_2 emissions until 2020. Moreover, the Scientific Advisory Council to the German Ministry of Finance suggested even to reconsider Germany's and the EU's ambitious Climate Protection Policies. Without an international binding agreement, most nations will do less resulting in economic costs, loss of jobs, less consumption, reduced growth rates etc. One must remember that every country acts to protect its national interests and has to respond to specific national background conditions; there is one fundamental reservation in comparing Japan and Germany: While the EU Members States have been pursuing their economic interests, they are trimming their economy more in order to converge their economic security with environmental security.

The situation is different in Japan. Thus far, Japan has preferred its economic fundamentals to environment. This can be clearly demonstrated through Japan's policy initiatives and statements and role during the Kyoto process, which conveys that it will extensively make use of the flexible mechanisms through the transfer of huge amounts of money for loans and the transfer of environmental technology to developing countries, which has the positive effect of securing its position in a growing market. An impetus on domestic efforts for reducing greenhouse gas emissions is missing as it might put a risk at the recovery of Japan's economy.

The dispute about the Whaling Convention and the Kyoto Protocol exhibits three different aspects relevant for this study:

1. The tendency through increasing transnationalization to prescribe certain sights, which results in major conflicts. These conflicts occur out of the lacking consideration and scrutinizing of different interests, so that they end up in opposition against a majority or a supposed hegemon.
2. The massive possibilities to influence the decision of governments from abroad, especially when, they are not taking a clear position against a certain policy, because
3. The environmental technology at the interface between economic interests and ecological considerations has a special role, which remains to be clarified in the further process of this study.

8.2 Bilateral

In 1959 it came to the first bilateral environmental treaty of the Federal Republic of Germany (FRG) with a neighbouring country in the broader sense. In the Agreement to Fight Animal Epidemics in the Border Area, Germany and the Netherlands agreed on joint efforts (Bundesanzeiger (Banz) 244/59) to fight disease that was spreading through the extensive transnational trade of cattle. Moreover, in 1961 the FRG and France bilaterally agreed on the Protocol Concerning the Constitution of an International Commission for the Protection of the Saar River Against Pollution (Bundestag-Drucksachen 12/5446: p. 6,617) to secure that the water supply was of adequate quality for the population and industries along the Saar River. In the following years until now, Germany has signed numerous bilateral environmental agreements, many of them very concrete in nature, while others have been rather generally targeting cooperation in the environmental protection (Auswärtiges Amt 1998).

Just as the FRG signed bilateral environmental agreements with their neighbours in West the German Democratic Republic (GDR) signed those predominately with neighbouring nation-states in the East. In 1950 and 1954 the GDR and the People's Republic of Poland signed the treaty on the Cooperation in Plant Protection and the Protocol on the Protection of the Fish in the Oder Bay, which can be considered as the first bilateral agreements of this kind (Bundesgesetzblatt (BGB) II 1993: p. 1,080; Bundesarchiv DK 5 VA 3,741).[85] These treaties of the GDR were followed by the agreement on the Peaceful Usage of Nuclear Energy (BGB II 1991: p. 1,077) with Czechoslovakia and the "Basis-of-Relations Treaty"[86] between the FRG and GDR in 1972, specifying environmental protection in Article 7 (BGB II 1973: p. 421ff).[87] However, within the context of unification of the FRG and the GDR and thus the unification contract (BGB I 1990: p. 885), all environmental agreements of the GDR expired, whereas those of FRG remained effective or were modified through negotiations.[88]

The rapid expansion and large catches of the Japanese high seas driftnet fisheries for salmon in the early 1950s helped lead to a treaty between Canada, Japan and the United States establishing a provisional Eastern boundary line in 1954 for the Japanese high seas salmon fisheries in the North Pacific Ocean at 175°W longitude and requiring the scientific determination of which line would best divide Asian and North American salmon in any areas of intermingling. However, Japan's first bilateral environmental agreement came only into force in 1956; this was concluded with the Soviet Union and aimed at regulating Japanese salmon driftnet vessels fishing on Soviet stocks. Further agreements on fisheries came into force in 1965 with the Republic of Korea, the People's Republic of China in 1975 and New Zealand in 1978.

There were further bilateral environmental agreements with neighbouring countries: The Agreement between Japan and the US on cooperation in the field of environmental protection (1975), the China-Japan Agreement for the Protec-

---

85 Schreiber (1986) describes the bilateral transnational environmental problems between the People's Republic of Poland and the German Democratic Republic in more depth.
86 German: Grundlagenvertrag.
87 The documentation of bilateral treaties, especially in the starting time of the GDR, is rather insufficient due to lax publication practice. Although this study incorporates the available documents of Germany's Federal Archive in Berlin, it cannot be ruled out that further treaties were existing, but are not detectable.
88 This is the result of comprehensive studies of the relevant original documents of the GDR put into the archives of the German Federal Archive in Berlin in comparison to legislative changes published in the Bundesgesetzblatt. As a follow-up to this result, the German Federal Ministry of Environment conducted a detailed analysis of the bilateral environmental treaties of the GDR and confirmed it later on; the results have not yet been published.

tion of Migratory Birds (1981), the Japan-USSR Agreement on Fishery Cooperation (1985), the Agreement Between the Government of Japan and the Government of the Republic of Korea on Cooperation in the Field of Environmental Protection (1993), Japan-China Agreement on Cooperation in the Field of Environmental Protection (1994) and the Japan-German Agreement on the Cooperation in the Field of Environmental Protection (1997).[89] Consequently, Japan concluded only few bilateral environmental treaties in the narrower sense, most of them dealing with the environmental medium sea and its flora. As indicated by the names of the above-mentioned treaties, the chief aims of these agreements were especially channelling the commercial fishing industry in order to ensure security of national interest in certain harvest quota. Certain long-term bilateral approaches that aim to protect the flora and fauna of the seas have been of less importance or do not have the necessary lobby support in Japan. In recent times bilateral environmental agreements have not been on the political agenda, simply because environmental problems were regarded predominately as "home-made" and thus left to solve at the domestic level. Yet extreme weather conditions, acid rain and its subsequent adverse ecological affects on the forests of the West coast of Japan, the increasing shrinking of fish stocks due to overfishing and the occurrence of BSE through international trade demonstrated that even island states are increasingly being affected by transnational environmental problems calling for appropriate political answers.

The asymmetric structure of interests between industrialized and industrializing countries is not necessarily an obstacle for a close bilateral cooperation between countries. It seems that the interests modify above all during different learning processes supported through various parameters.

---

[89] With Korea: http://www.mofa.go.jp/policy/economy/apec/1995/member/info/9.html – 03 October 2010.
With China: http://www.mofa.go.jp/policy/economy/apec/1995/member/info/5.html – 03 October 2010.
With theUSSR: http://www.earthtrust.org/dnpaper/history.html – 03 October 2010.

9 Framework Conditions

A critical analysis and evaluation of Japan's transnational environmental policies in the 1990s requires an investigation of the framework conditions within these policies developed. The following comparative excurse some basic data and history will not only help represent the relevance of such for the development of transnational environmental policies, but also impart necessary information of circumstances and practices, which among others Flyvbjerg (2001) also considered as important for a comparative analysis of the overall performance of Japan. They are a part of the set of interactions that influenced the political results under analysis here and will assist in verifying or falsifying Dehnhard's theory.

9.1 Impact on the Environment

Despite its small stake in world population and area, Japan can be categorized today as an economic superpower.

Japan's GDP in 1998 was US$ 2,543.8 billion (12.77%) comprising almost one-eighth of the OECD GDP in 1998[90]. Until 2008 Japan's GDP has increased to US$ 4,910 billion. These figures include expenditures for curative measures against environmental pollution and damages through natural catastrophes, which are finally also identified as economic yields.

Table 3: Gross Domestic Product – Japan, USA, Germany (2002 & 2008)

Country	GDP in 2002/2008 (in bUS$)	% of total worldwide GDP in 2002/2008
Japan	3,918/4,910	11,84/8,10
USA	10,417/14,093	31.5/23,27
Germany	2,016/3,649	6.1/6.02
WORLD	33,070/60,557	100

(Source: The World Bank 2010)

9.1.1 Imports of Forest & Agriculture Products

Although rich in forest resources, Table 4 shows that Japan (US$ 17,184 billion) was in total numbers the second largest importer of forest industry products be-

---

[90] Germany US$ 3,649 billion (6.02%) + Japan US$ 4,910 billion (8.1%) = US$ 8,559 billion (14.12%) – comprising one-seventh of the world GDP in 2008 (The World Bank 2010).

hind the US (US$ 24,120 billion) but before Germany (US$ 12,127 billion) which was placed third in this ranking (OECD 1999: p. 130). Timber is largely used as a raw material in forest industries that process saw wood and sleepers, wood-based panels, wood pulp, and paper and paperboard, playing an important role in both Japan and Germany. However, in sharp contrast to Germany and the US, who have been campaigning against imports of tropical timber, Japan was in total amounts the number one importer of cork and wood from tropical countries, albeit on per-capita basis with US$ 9.1/cap – far behind Portugal (US$ 19.9/cap), Belgium (US$ 16.5/cap) and Korea (US$ 11.1/cap) in the 1990s. The harvesting of wood from tropical forests and the export of a part of this harvest is still today a significant cause of tropical deforestation. Tropical wood harvesting also has an impact on the world's genetic resources and the increasing CO_2 concentration in the atmosphere. Japan's total amount of imports of cork and wood from tropical countries, however, decreased from US$ 2.697 billion in 1995 to US$ 1.145 billion in 1998 (OECD 1999: p. 134), which is equal to a 58% decrease within three years, but nevertheless remaining at the top of the list. One important reason for this reduction, as explained by Japanese officials, was the fact that one-way chopsticks are now predominately produced with domestic timber.

Table 4: Timber Imports – Japan, USA, Germany (1998)

Country	Import of industrial timber products (in bUS$)	Import of tropical timber (in bUS$)	Total per-capita (US$/cap.)
Japan	17.184	1.145	9.1
USA	24.120	0.426	1.6
Germany	12.127	0.157	1.9
Total OECD	113.001	5.586	5.1
WORLD	145.477	N/A	N/A

(Source: OECD 1999)[91]

To satisfy the needs of agricultural products that Japan and Germany consume throughout one year, both nations have rather high import quota as illustrated in Table 5. Already in 1992 the Japan National Institute for Environmental Studies (NIES) undertook a test calculation to demonstrate Japan's reliance upon the soil of other countries. This study came to result that about 10.3 million hectares of land was taken up all over the world for the purpose of producing

---

91   The latest OECD Environment Data Compendium (2004) does not include any information on tropical timber imports.

wheat, barley, sorghum, maize, soybeans, coffee beans, cotton and natural rubber for Japan. This was an area equivalent to 2.3 times the total cultivated land area in Japan and about 27% of the nation's total land area (Environmental Agency 1995: p. 155). Moreover, declining land utilization rates and abandoned cultivated land is resulting in inadequate soil management in Japan. Consequently, in 1999 Japan imported 11.4% of the total amount of worldwide imports (US$ 308 billion) of agriculture products (FAO 2000) – all EU countries 17.8%.

Table 5: Trade in Forest Industry Products – Japan, USA, Germany (2002)

Country	Imports (in bUS$)	Exports (in bUS$)	Imp-Exp Balance (in bUS$)
Japan	10.480	0.920	-9.6
USA	23.353	13.828	-9.5
Germany	11.108	11.165	0.057
Total OECD	106.727	103.098	-3.629
WORLD	139.798	133.225	6.573

(Source: OECD 2004)

Although it is an agricultural products exporting country, Germany imported 3.4 times more of those products than it exported in 2001 (Federal Statistical Office of Germany 2003). About 36% of Germany's total area is arable acreage, 16% pasture. However, in contrast to Japan, Germany is in the position to produce a large part of its food essentials – up to 80%.[92] Germany also does not have to import either meat or dairy products, but through transnational trade under international trade agreements and especially through intra-EU trading, these meat and/or dairy products are increasingly appearing on the German market.

This became apparent by October 1996, when BSE[93] had been first reported from 10 countries and areas outside the United Kingdom including Germany and the reappearance of the Foot-and-Mouth-Disease (FMD) in February 2001

---

92 See http://www.straubing.baynet.de/~k.czauderna/BRD.htm – 23 August 2004.
93 Bovine Spongiform Encephalopathy (BSE) first came to the attention of the scientific community in November 1986 with the appearance in cattle of a newly-recognized form of neurological disease in the UK. Between November 1986 and May 1996 approximately 160, 000 cases of this newly-recognized cattle disease were confirmed in the UK. Epidemiological studies conducted in the UK suggested that the source of the disease was cattle feed prepared from carcasses of dead ruminants and those changes in the process of preparing cattle feed introduced in 1981-1982 may have been a contributing risk factor. A newly recognized variant of the Creutzfeldt-Jakob Disease affecting human brains is expected to be linked to BSE (WHO 1996).

in the UK. Japan, being the largest importer of meat (28%) in the world, identified its first case of BSE in Chiba Prefecture in September 2001. As demonstrated by this case, excess dependence for both exporting and importing countries is causing problems with regional effects that are capable of crossing national borders and thus must also be considered as transnational environmental problems.

However, the dependence from agricultural products of other countries describes only one facet of Japan's lack in raw materials.

9.1.2   Imports of Resources

The environmental impact of industrial activities may vary considerably with structural changes, as experienced by Japan in recent years. The decline of the so-called traditional industries e.g. such as iron and steel has reduced quantitative pressures on the environment in this environmentally significant industrial sector. This downturn was accompanied with an increase of the total production in the electrical machinery industry (9%), petroleum refineries (4%) and motor vehicles (3%); surprisingly, Germany's iron and steel industry maintained its production level from 1995 but also in 1998 due to subsidiary support from the government. Considerable increase can be recognized in the chemical product industry (20%), electrical machinery (12%) and motor vehicle industry (27%) (OECD 1999: p. 255f). The growth in these new industries can create not only qualitative pressures, such as soil and groundwater contamination by trace toxic chemicals, but also requires immense imports of necessary resources.

Modern ICT equipment plays a crucial role in today's economies and societies; worldwide sales in the ICT sector in 2002 totalled US$ 1,104 billion, well on the way to crossing the US$ 1,400 billion mark in 2003 (EITO 2002: p. 391). Of the total amount of ICT, 41.7% was generated in the US, 13.1% in Japan, and 6.7% in Germany contributing substantially to the real economic growth in the both nations under comparison. On a per-unit basis, a study by Atlantic Consulting (1998) for the European Union estimated that the final production of a personal computer (including material production, manufacture and distribution) would lead to the release of 0.19 metric tonnes of greenhouse gases, 36 kg of overall waste and would require 3.6 GJ of energy. Consequently, ICT use also has implications for resource inputs other than energy. From an environmental perspective, those garnering the most significance are the metal and materials sectors. While there is a continued migration towards plastic, rather than metal parts, the critical functions of this type of equipment are achieved by advances in uses of metals and materials; metals still used in ICT equipment are aluminium, lead, cadmium and nickel.

Table 6 illustrates the imports of a selection of raw materials typical of the Japanese and German industry in the 1990s. The mining of coal and oil, their transport and also their use generates pressure on the environment. This includes pollution of air, water and land, consumption of natural resources for production chains, negative effects on wildlife and natural areas and spills from production and transport. Similar impacts are associated with the mining of phosphate rock, which is imported to produce fertilizers and animal feed supplements and the mining of salt for consumption and industrial processes. The use of natural rubber as biopolymer is widespread, ranging from household to industrial products with tires and tubes as largest consumers. A critical issue in the natural rubber production is the threat through millions of hectares of rubber plantation monocultures. Iron ore are rocks and minerals from which metallic iron can be economically extracted. Iron is the world's most commonly used metal; steel, of which iron is the key ingredient, is used in engineering applications, the automotive and maritime sector as well as the machinery industry. The mining of iron ore is a high-volume low-margin business with environmental impacts similar to the mining of the above-described raw materials.

Table 6: Imports of Selected Raw Materials – Japan & Germany (1997 & 1998)

Raw Material	Japan		Germany	
	Quantity (in 1,000t)	Value (in mill. US$)*	Quantity (in 1,000t)	Value (in mill. US$)**
Coal	131,764	8,010	24,565[1]	847[1]
Salt	7,914	290	1,705	34
Phosphate Rock	976	110	223	11.3
Oil[a]	280,300	--	137,400	--
Iron Ore	120,782	3,980	47,597	1,231
Natural Rubber	669	740	345	239

* Calculation on the basis of JP¥ 100 = US$ 1
** Calculation on the basis of DM 2 = US$ 1
[a] Net oil imports in 1997, source: OECD 1999
[1] lignite + hard coal

(Composed out of data of OECD (1999); Japan Statistical Yearbook (2000) & Federal Statistical Office Germany (2000))

9.1.2.1 Case of Aluminium

In 2000 the imports of aluminium and aluminium scraps amounted 2.473 million tonnes in Japan (Japan Statistical Yearbook 2003) and 1.090 million tonnes in Germany (Drasdo 2001: p. 10), most of those from countries with relatively lax environmental legislation such as Russia or Indonesia, but also Australia. Aluminium ore, most commonly in the form of bauxite, exists mainly in tropical and subtropical areas. Bauxite is refined into aluminium oxide trihydrate (alumina) and then electrolytically reduced into metallic aluminium. Bauxite is generally extracted by open cast mining from strata, typically some 4-6 metres thick under a shallow covering of topsoil and vegetation. In most cases the topsoil is removed and stored having an enormous impact on the landscape.

The production of primary aluminium uses a great deal of electricity. The environmental profile for this part of the aluminium's life cycle is therefore highly dependent on how the electricity use is generated. For example, Germany's aluminium production on the basis of lignite and coal amounts already to 15 kg CO_2 and 45 kWh of primary energy per kilogram of aluminium. In environmental terms the most important emissions from the aluminium production process are: (i) the emission of CO_2 and CFC – greenhouse gases, (ii) SO_2, which contributes to acidification and (iii) fluorides and polyaromatic hydrocarbons (PAH compounds), which are toxic to humans and the environment. In response to these environmental impacts, Japan's former Ministry of International Trade and Industry (MITI) has proposed to build an aluminium smelter overseas. In Germany governmental officials are continuing to emphasize the importance of coal and lignite – making up 33% of Germany's energy production – in order to secure Germany's energy demands.

9.1.3 Greenhouse Gas Emissions

Since the beginning of the 1980s, a large percentage of Japanese industry has been manufacturing and storing its products in other Asian countries. The aim has been to benefit from the strength of the yen, the low storage costs and most importantly, the less rigid environmental controls in other countries. With the implementation of Germany's law on pollutants[94] in 1986, which e.g. resulted in investments of EUR 10 billion of Germany's energy producers in order to install effective filter-systems in their power plants using fossil resources, many German companies shifted parts of their production to other countries; currently, Germany's metal producers are only using high-concentrated pre-processed ores.

---

94 Bundes-Immissionsschutzgesetzes (BImSchG).

Table 7: Emissions of Greenhouse Gases, NO_X and SO_X – Japan, USA, Germany (2002)

Country	Greenhouse Gases					NO_X (mt)	SO_X (mt)
	CH_4 (mt)	N_2O (mt)	CO_2				
			from energy use (mt)	from industrial processes (mt)	total (mt)		
Japan	19,5	35.4	1,178	249.3	1,331	2.02	0.86
USA	598,1	415.9	5,705	621.4	6,934	18.83	13.85
Germany	81,4	55.8	848	124.2	1,014	1.42	0.61
Total OECD	1,285.4	934.6	12,600	1,833.1	14,472	39.5	32.1
WORLD	--	--	23,710	4,322.9		--	--

in one million tonnes (mt)
(Composed out of data of OECD (2004), (1999) & (1997))

Further examples for Japan's impact on the transnational environment and also for its transnational environmental responsibility result from statistics on emissions of gases. Table 7 illustrates the man-made emissions of nitrogen oxides (NOx) and sulphur oxides (SOx). This table refers to the major categories of emission sources of these pollutants, which include motor vehicles, power stations, fuel combustion, pollutants emitted in manufacturing, waste incineration and agricultural burning etc. Japan holds the second lowest place in a direct comparison of per-capita emissions of NOx from OECD countries and is behind Norway and the Netherlands in regard to its SOx emissions; Germany is also more than 50% under the OECD average. One should note that these emissions of traditional pollutants result not only from strict domestic environmental legislations, but also from the shifting of certain high-polluting industries to outside the country.

Behind the US and the Russian Federation, Japan and Germany were taking up place three and four in the hierarchy of top greenhouse gases emitters in the 1990s (OECD 1999: p. 53). Greenhouse gases disturb the balance of the earth's radiation energy budget which leads to an increase in the earth's surface temperature and to related effects on e.g. the climate, sea level, desertification etc. CO_2 is the gas that contributes the largest share, most resulting from energy use. Japan's CO_2 emissions from industrial processes are not much smaller than that of the US, but Germany's emissions are less than half of that of the US and Japan.

The above-given examples are a few of the many indicators demonstrating Japan's impact on the transnational environment as an economic and industrial world power. This results not only from industrial production processes, but also from consumption manners which are unsustainable in many ways and thus, also having an impact through e.g. the exploitation of natural resources, the production of metals, the extensive burning of fossil fuels for energy production and certainly the decreasing life spans of many products. All of this extensively influences not only the global environment, but also local/regional environments far away from Japan.

The above given examples for Japan's demand on the Earth ecosystem through the imports of resources such as oil, coal and iron ore and the resources and energy-intensive production of e.g. aluminium demonstrate that some interstate interactions may involve governments, but also other actors. Industrial, financial, international and non-governmental actors play a significant role here; consequently, interstate interactions are considered "transnational" (Menderhause 1969). Based on the growing importance of non-governmental actors in the field of international environmental policies and development policies,[95] and especially also in the case of environmental technology transfer through public-private-partnerships (PPP), it appears justified to speak of "transnationalization". Transnationalization is characterized through the growing role individuals and organizations play in world politics vis-à-vis foreign government or foreign societies and thus bypass their own governments (Singer 1969: p. 24). It has become clear that globalization and transnationalization processes perforate border lines drawn by nation-states, which finally leads to a polycentric distribution of power addressed by George *Rosenau* in 1990. This points to a turning away from the sovereignty of Hobbes's Leviathan and Herz's hard shell, which understand the statesperson as the only source of governance represented through a government (Hobbes 1996: p. 144; Herz 1976: p. 100). Additionally, it also reflects a key error in current globalization debates: the identification of the modern state with the nation-state. Were there any indicators in Japan's transnational environmental policies in the 1990s that this transnationalization necessarily led to a constant loss of control and governance ability of states, making its claim for "being a flagship" ridiculous?

---

95 See e.g. Weiß 2000.

Part C – Analysis

Whereas Part B of this study was on the history of transnational environmental cooperation of Japan, Part C provides the actual analysis of Japan's policy in the 1990s. This is done on the basis of the Policy-Cycle Model (PCM) and is therefore in line with the requirements of Dehnhard's theory.

10 Initiation

The main requirement for eco-political steering through transnational policies is the perception of an environmental problem. For subsequent action the first step is to identify all information required, raised and transferred by different social groups, individuals and institutions, so that it moves into the public consciousness. When political protagonists recognize the call for acting, the problem is set on the political agenda for decisions (Windhoff-Héritier 1987: p. 65). This first step of the first phase of the PCM is designated as perception of a problem or initiation.

10.1 Development of Transnational Relations

The arrival of Americans and Europeans in the 1850s increased Japan's domestic tensions. The Tokugawa *bakufu*, or shogunate, already weakened by an eroding economic base and ossified political structure, found itself challenged by Western powers. Their intent was an opening of Japan for trade and foreign intercourses and the deliberate and successful isolation of Japan during the Tokugawa period from 1639 was ending. This isolation was not only favoured through its offside location, but also through its lack of desirable natural resources.[96] Furthermore, it was favoured through the self-centredness of China, the confucianistic contempt of warriors and the lacking maritime ambition of this gigantic empire. All of this supported the cultural and political autonomy of Japan but this predominate inland orientation for more than 200 hundred years was cracked through a forced opening of Japanese harbours by Commodore Matthew C. *Perry* in 1854; this was followed by openings of various harbours for European nations, the signing of contracts, and providing the great powers looking for potentially new markets with enormous rights in Japan resulting in

---

96 Only the Chinese and Dutch were allowed to call at Nagasaki harbour, which was under control of the Japanese government in Edo (today's Tôkyô). For more details see e.g. Bersihand 1963, Dettmer 1965, Hammitzsch 1981.

the erosion Japan's sovereignty. The implications of such contracts and contacts were unclear to the Japanese who were rather inexperienced transnationally and internationally and finally resulted in conflicts between foreigners and the Japanese.

This time of upheaval characterized through the weakness of the Tokugawa state in responding to the threats to colonize Japan was followed by domestic military conflicts between members of feudal military clans and the troupes of the *Shôgun*, which finally lead to the overthrow of the last Tokugawa-Shôgun in 1867. With the return of power to the imperial house in 1868 by emperor Mutsuhito, later known by his reign name Meiji (inspirited governance), enormous changes initiated through the revolutionizing feudal forces, which laid the foundations for a development of a modern Japanese economy and society (Berger 1987). These reforms came to an end in 1889 through the proclamation of a constituency following the Prussian model. The feudal military caste linked with the traditional families of traders, from which a new world of banks, industry and trade developed headed by the Japanese family groups named *Zaibatsu*. The Japanese government financed model firms to ensure fast growing industrialization, by which the Japanese learned about western techniques and organizations. Max *Weber* considers this as one of seven ideal pre-requisites for the building of a private-capitalistic industrial revolution.[97] The other pre-requisites are:

- Free acquisition of means of production,
- Free trade and free markets,
- Rational right,
- Free labour,
- Commercialized economy,
- Free trade of enterprise shares and fortunes.

These did practically not exist in Japan until 1868; consequently, the industrial revolution began about 100 years later than in Europe with the publicly owned textile industry as a model. The import of Western technologies and the exports of Japanese textile products were important currency earners.

With growing financial resources the main emphasis was given to heavy industry to build a powerful army and fleet as the inferiority to the Western military was a big shock to the Japanese. However, at the sight of its naturally given maritimity and supported through an enormous improvement and expansion of worldwide shipping due to expanding worldwide trade and supported through new techniques, a growing motorization and higher safety, further essential prerequisites for Japan's development, were given. Thus, the Japanese islands be-

---

97 See Weber 1991.

came central ports half the way from the Southeast Asia to the Pacific Coast of the US.

At the beginning of industrialization, Japan and Germany developed out of rather different framework conditions.[98] The contacts and the knowledge of developments and the situation in other countries were an important driving force to catch up the economic backwardness and thus to form the basis of an own big industry. Both states were motivated to regain power in the international sphere through industrialization.

Japan extensively adopted Western sciences and techniques which finally led to an astonishing reception of foreign ideas. Germany actually contributed to the decrease of Japanese discoveries and patents by developing its own. Thus the transnational contacts of Japan proceeded just after its opening in essentially one direction – from outside into Japan, whereas Germany's contacts proceeded in both directions across its borders.

---

98 Already more than 50 years before Japan, on the threshold of the 19th century, Germany experienced far-reaching social, economic and political reforms, creating a foundation for the performance of today's society – with all its fascinations, but also unmistakable problems (Weismantel 1987: p. 172). Especially the sharp rise of Germany's population at the end of the 18th century must be considered as a green light for an industrial society following the new examples for entrepreneurship in England and France. Germany's economy had been controlled by aristocrats, who turned their nose up at profitable activities. However; the Prussian reform laws in 1807 and 1810 increased regional and social mobility and small companies were discharged from the cramped corset of classes. New initiatives and entrepreneurial spirit were needed to solve the domestic challenges, but also to keep abreast at England. The embargo to import goods and natural resources from England and thus its colonies as a consequence of the Napoleonic war contributed to the blossoming of German industry but with the lift of the blockade, the sham boom of Germany's industry shattered through products hitting from England. The 1850s, characterized by the barren politics and the German confederation as such, rigid and unyielding after the complete crush of the revolution in summer 1849, was fully re-established remained blind to the need for reform. These years were of the highest importance in economics, as the period of great breakthrough of industrial capitalism occurred during this time in central Europe. The national energies, frustrated in the effort to achieve civic reform, turned to the attainment to increase the standard of living through material progress. This concentration of forces was followed by an economic expansion and encouraged speculative boom so that new investment banks were founded to provide risk capital for factories and railroads, the basic transportation system for Germany at the time. Although the rural population still outnumbered the urban, the tendency toward industrialization and urbanization had become irreversible and as wealth continued the shift from farming to manufacturing. The pressure toward a redistribution of political power also gained strength.

10.2 Phase of Ecological Ignorance

In the years of rapid economic growth after the end of the Second World War, economic growth in Japan enjoyed supreme priority. The national economic policy decisively concentrated on heavy industry and chemical organizations (ICETT 1994a: p. 3). An era of mass production had begun, in which black smoke of the chimneys of industrial plants became symbols of prosperity and not for the destruction of the environment. In the mid-1960s heavy oil became the major source of energy. With the expanding industry the problem of the emission of sulphuric acid gases in the atmosphere increased.

Just two years after the start of the petrochemical industry at the bay of Ise near Yokkaichi (prefecture of Mie), respiratory tract diseases became first apparent in form of e.g. asthma. At the beginning of the 1960s, epidemiological investigations already pointed to the health threat through the petrochemical industry there. Still, responsible bodies failed to introduce effective measures for the protection of the environment and the enterprises capacities were expanded further. In 1972, after a five-year lawsuit, the civil courts met the compensation demands of victims of the "Yokkaichi-asthma" and its equivalent and the managers of the petrochemical industry of this region were found guilty (Imura 1993: p. 65).

The Minamata Diseases became known in the public after the discovery of one unknown cerebral disturbance in a child in 1956. The comprehensive investigation through the Kumamoto University in 1959 endorsed the conjectures that mercury, released by enterprises in the bay of Minamata (Prefecture of Kumamoto), caused this disease.

Pathologists discovered a high mercury concentration in tissue samples of fish and fishermen of this region (Ui 1992: p. 110). Concrete measures on the part of state institutions under application of the Public Water Zone Conservation Law and the Industrial Effluent Law to pinpoint the reason for this mercury contamination in the bay of Minamata, however failed to materialize. Similar diseases appeared later on in June 1965 along the Aganogawa River in the prefecture of Nigata, frequently leading to death, like in Minamata. Again, a mercury combination was supposed to have caused this disease and again, initiatives for a lasting solution to this problem failed to coalesce. It was only in 1967 that the first basic environmental law was enacted; for the time being it contained a paragraph prescribing that the protection of the environment should take place in harmony with healthy economic development (Kelley et al 1976: p. 244). This harmony was often referred to by the authorities for allowing economic aspects to take precedence over environmental ones. The environmental load shifted from rather limited local areas in the countryside to other areas in the country, including urban areas. In 1999 in an interview with the author, Jun *Ui* described these developments as a path towards ecological *harakiri*.

Hereafter one family decided to demand compensation through legal channels and simultaneously to clarify the responsibilities for this pollution. Resulting from this was a citizens' movement of victims of Minamata Disease and of Itai-Itai-Disease in the prefecture of Toyama (cadmium contamination with effect on bone formation); today, this group of lawyers, doctors, and scientists still represents the interests of environment victims. This citizens' movement was successful in winning four large environmental civil lawsuits where high compensation payments from the accused enterprises were paid (Weidner 1988: p. 145).

In the literature the influences of the citizen movements, the opposition parties, the city administrations, the prefecture governments and the courts admitting the guilt of the state government have been analyzed sufficiently for not recognizing the massive pollution of the 1950s and 1960s and subsequently for not reacting (e.g. McKean 1981; Reed 1986; Tsuru and Weidner 1989). These investigations denoted that this citizens' movement, along with its pluralistic interests and the jurisdiction of the courts, prompted the governing LDP party (Liberally Democratic Party) to put pollution on the political agenda, primarily because the governing party continually lost votes in their rural strongholds.

The government reacted to this development in a special session with the decree of fourteen laws and ordinances on environmental protection as well as measures for the reduction of unhealthy pollutant emissions. The clause on harmony between environmental protection and economic development was omitted from the basic environment law. The repressive – at best symbolic – Japanese environmental policy of the 1950s and 1960s (Weidner, Rehbinder & Sprenger 1990: p. 35; Weidner 1996: p. 158), which refused minimal ecological responsibility considering an economy purely orientated to expansion, followed an eco-political turn towards a more technocratic and active environmental policy (Weidner 1996: p. 160).

More severe environmental laws were enacted and realized primarily to rehabilitate existing domestic problem situations. The Environmental Agency was established in 1971 and such developments also indirectly affected preventative actions by private and public enterprises.

Despite this turn ecological depletion continued in Japan, mainly through shifting production to Southeast Asia but also because of the lack of environmental laws (Maull 1992: p. 362ff.).

Likewise, as Japan's foreign policy is counted as Janus-faced (Kevenhörster 1993a: p. 16), its environmental policy demonstrates the same in the 1970s and continuing until the late 1980s. The one face showed the successful efforts for an effective environmental policy in their own country, the other face the massive exploitation of nature, predominantly in countries of the Asian-Pacific region through Japanese enterprises.[99]

In contrast to Japan, but typically for the industrializing countries in Europe, first regulations against water and air pollution were already enacted in Germany in the 19th century. According to Article 15 of the Weimar Constitution environmental protection was one of its objectives in 1919. And in 1957 the Federal Republic of Germany (FRG) introduced its first water management law[100] (Hünemörder 2004). For a long time environmental considerations were not acknowledged in Germany, despite the obvious air pollution in Europe's largest industrial region, the Ruhr Area, through the burning of coal and lignite, metalworking and chemical industry. Companies like Krupp and Thyssen were essential for Germany's armaments industry but diseases like typhus and cholera, a result of the pollution of water with ammoniac, chlorine, saltpetre etc, were part of daily life.

During the Cold War the Ruhr Area's heavy industry was regarded essential for Europe but with the early 1960s air pollution began to constitute alarming proportions, especially during the winter smog of 1962. Diseases such as leukaemia, cancer, rachitic and changes in blood count were exploding, substantially affecting children. The death rate alarmingly increased but the chances to go to court because of health affects resulting from pollution were minimal for a long time. Scientific work identifying the responsible parties for harm to health and environment were lacking; customary local contaminations were regarded acceptable.[101]

The situation in the Ruhr Area in the early 1960s illustrates the force of economic arguments against environmental; the main values were jobs, tax income and prosperity and the power with governments and industry. Civil movements against the exploitation of the environment were shy and governments, industry and public administration could only agree on the necessary for direct protective measures. In consequence in the same way as in Japan, the FRG's policy in the early 1960s ignored environmental responsibility, mainly following economic motifs. However, with Willy Brandt's[102] campaign to "blue the sky over the

---

99   For example the high amount of tropical timber from the Southeast Asian rainforest imported by Japan.
100 In German: Wasserhaushaltsgesetz.
101 For more details see e.g. http://www.route-industriekultur.de/sonstiges/daten-und-fakten/facetten-der-region/der-blaue-himmel-ueber-der-ruhr.html – 23 July 2010.
102 Willy Brandt, was a German politician, Chancellor of the FRG 1969–1974, and leader of the Social Democratic Party of Germany (SPD) 1964–1987. Brandt's most important legacy was a policy aimed at improving relations with East Germany, Poland, and the

Ruhr Area" in 1961 and with an environmental civil movement gaining momentum, the 1960s also indicate the turning point in the FRG's environmental considerations, leading to first countermeasures against smoke and soot through e.g. higher chimneys – finally resulting in substantial problems with acid rain in the 1980s. Also the media-campaigns that addressed the increasing pollution of the Rhine River through chemical releases resulting in the depletion of millions of fish led to a public outcry (Hünemörder 2004: p. 113ff). Finally, with the new federal government of Willy *Brandt* and Walter *Scheel* (1969-1972), the policy field "environment" was introduced. This was also initiated by the environmental movements in the US and the preparation for the UN Environment Conference in Stockholm in 1972 and resulted in the ad hoc programme in 1970; one year later it generated a detailed environmental programme including a roadmap for necessary legislative measures (Jänicke 2009).

In 1974/75 the oil crisis resulted in a certain political countermovement due to a recessive economic development. The civil movement was now actively demanding the realization of the ambitious environmental targets through the new government under Helmut *Schmidt* (1974-1982) which were set already under Willy *Brandt*.

The environmental policy in the former German Democratic Republic (GDR) is not to be ignored here. In 1968 the GDR constituency incorporated the protection of the environment as a duty for both the state and society. The following constituency norm (*Verfassungsnorm*) resulted in a national cultural law (*Landeskulturgesetz*) principally regulating the GDR's environmental policy. Moreover, the GDR was among the first establishing a Ministry of Environment and Water Management in 1971. Nevertheless, these steps do not abrogate the disastrous ecological problems through air and water pollution in the GDR. The regime in East Berlin was giving the industrial development absolute priority over environmental protection. Only in the 1970s and 1980s the GDR was confronted with the highest sulphur dioxide pollution in Europe due to the extensive burning of lignite; only the waste problem was less severe in comparison to the FRG and other industrialized nations. The substantial lack of resources and packing material resulted in a substantive utilization of municipal and industrial waste but overall, environmental problems were regarded as characteristic for the capitalistic world following imperialism. Civil movement against environmental problems in the GDR were branded as opposition against the SED re-

---

Soviet Union. This policy caused considerable controversy in West Germany, but won Brandt the Nobel Peace Prize in 1971.

In 1974 Brandt resigned as Chancellor after Günter Guillaume, one of his closest aides, was exposed as an agent of the Stasi, the East German secret police.

gime[103] and participants of these civil movements were persecuted. In consequence the history of environmental policy in the GDR also illustrates an ecological ignorance as already explained for the FRG and Japan.

In contrast to Japan, which enacted is first law to protect the air in 1968, the FRG followed only in 1974 and continued with the establishment of an Environmental Agency[104] and increasing research around environmental problems. Also in contrast to Japan, the FRG's environmental policy in the 1960s and 1970s cannot be categorized as Janus-faced, because for a long time it did not substantially move its production industry to neighbouring countries without any or only lax environmental regulations. Even so in both Japan and Germany economic considerations prevailed supported by a close alliance between industry and governments; Germany's first countermeasures only shifted the problem from one environmental sphere to another – typical for end-of-pipe or purely cleansing approaches. The transfer of production sites of Japanese companies to countries in the Asia-Pacific region was also an attempt to shift the issues outside the country and hence bypass costly countermeasures resulting out of the stricter domestic regulations. This step also resulted in an increased consideration of Japan's responsibility in its neighbouring countries, which was one dimension of its environmental technology transfer initiatives.

The above illustrated that beyond the government as a key national actor supported by public administration, industry was playing a key-role during this time of ecological ignorance. It was only with a growing civil movement that questioned the government's and industries' values of primarily only supporting the economic development and risking the governments' power, environmental considerations were introduced to the political agenda. These phases took place before a growing global environmental consciousness and transnationalization as described in Chapter 2.2. In consequence these interactions between governments, businesses/industries and the people do not necessarily portray any development towards a diminishing of the state element in governance, but the classical policy development process, in which the heads of states were for a long time only serving their own interests or those directly stabilizing their power more than often including businesses/industries.

---

103 The Socialist Unity Party of Germany (German: Sozialistische Einheitspartei Deutschlands, SED) was the governing party of the German Democratic Republic from its formation on 7 October 1949 until the elections of March 1990.
104 In German: Umweltbundesamt (UBA).

10.3 Growing Environmental Consciousness

The protection of environment and resources moved into the centre of international discussions in the 1970s (Nuscheler 1995: p. 247f). The UNCED in Rio de Janeiro (1992) (König 1995: p. 18), but also the Rio+5 Conference in New York (1997) and the WSSD in Johannesburg (2002) stressed the need for political action on the global environment problem.

After a phase of stagnation, especially at the end of Chancellor Helmut Schmidt's term, it was to the large surprise of most that the new FRG Christian Democrat liberal government under Helmut *Kohl* (1982-1998) started introducing massive air pollution controls through e.g. regulation of the large combustion plants[105] and the emissions control of automotives, making the FRG a European forerunner in environmental legislation (Jänicke 2009), but also a leader in an increased cooperation between government and industry. Moreover, environmental NGOs and the public had strengthened their pressure on companies and branches responsible for substantial environmental impact e.g. through the media etc. This development was supported by the success of the party "the Greens" during the general election in 1983 introducing environmental considerations into FRG's political and parliamentary system. The massive loss of forests through acid rain as a result of unsustainable countermeasures in the 1960 and 1970s helped in creating public awareness. On 6 June 1986, only six weeks after the accident in the nuclear power plant in Chernobyl (Ukraine),[106] the Kohl government decided to concentrate forces in a newly established Federal Ministry for the Environment, Natural Protection and Nuclear Safety (BMU) in order to take better steps against these environmental challenges. Before the creation of this new ministry, environmental aspects were dealt with in various ministries. Klaus *Töpfer*[107] following the Founding Minister Walter *Wallmann* guided the FRG's and subsequent unified Germany's entire environmental policy to substantially growing global consciousness in 1987 laying the groundwork for its environmental technology transfer in the 1990s as part of its system-immanent tendency towards ecological modernization; this is going to be described and analyzed later in this study. In addition to Chernobyl, a number of environmental catastrophes and increase in international activities in the late 1980s and early 1990s provided tail wind for these developments. Such exam-

---

105 In German: Großfeuerungsanlagenverordnung.
106 The Chernobyl accident is considered the worst nuclear power plant accident in history and is the only level 7 event on the International Nuclear Event Scale.
107 Executive Director of UNEP from 1996-2006.

ples include the Sandoz chemical spill[108] in November 1986, extensive dying of seals in the North Sea, the growing climate change discussion and the Report by the World Commission on Environment and Development, *Our Common Future*.

It was only at the end of the 1980s that the political protagonists of Japan saw relevance in environmental policy area for the second time after the tragic cases of mercury poisoning, cadmium poisoning and asthma during the 1950s and 1960s. Various representatives in Japan declared to take a leading role in their country on international environmental policy. This interest stands in sharp contrast to Japan's attitude prior to 1988; before 1988 the Japanese public, environmental groups, media and political parties paid little attention to the global environmental issues. Large US news magazines like *Business Week*, *Time* and *Newsweek* sharply criticized Japan's role as a global polluter thereby effectively referring to the contribution of Japanese enterprises to deforestation of the tropical rainforests through the worldwide supreme import quota of tropical timber, drift-net fishing, the trade with products of threatened animal and plant species as well as the shift of high-polluting industries to Southeast Asia (Subchapters 9.1.1; 9.1.2; 9.1.3; Maull 1992: p. 358). By 1988 neither a growing interest of the Japanese public in global environment problems was recorded nor was an evident transnational environmental problem apparent.

Up to this surprising change, Japan's international environmental policy was viewed as reactive due to the lacking research in the global environment area, the absence of strong eco-political representations and a weak environment authority (Environment Agency [EA]) (Schreurs 1994: p. 36f). Increasing political pressure from foreign countries through international environment organizations like Greenpeace and Friends of the Earth (Imamura 1989: p. 43ff), from internationally active enterprises confronted in foreign countries with environment themes and environment activists, and - perhaps the most important catalyst - the pressure of other governments became more effective when known Japanese politicians themselves could combine environmental involvement with their own interests. Simultaneously, the climate problem appeared threatening for the Japanese public because of increasingly hot summers with increasing drinking water problems in some regions and colder winters as well as heightened forest damage caused by emissions originating from the Chinese mainland (Park 1995: p. 32). For years Japan was criticized for undertaking insufficient international

---

108 During such toxic agrochemicals were released into the air and resulted in tonnes of pollutants entering the Rhine river turning it red. The chemicals lead to massive mortality of wildlife killing, among others, a large proportion of the European eel population in the Rhine river.

responsibility as an economical and technological giant; in the late-1980s some Japanese politicians adopted new images of statesmen with global concerns:

Prime Minister Noboru *Takeshita* adopted a more active foreign policy during his term in office (06 November 1987 - 02 June 1989) and especially as chairman of an International Environmental Conference in Tokyo in 1989, in which Japan promoted itself as a voice for developing countries and as a bridge between these and the (post-) industrialized world. During his term as Prime Minister, Takeshita showed only little interest in eco-political themes; however this LDP politician, surrounded by scandals, tried to polish his reputation regarding a second term in office as Prime Minister through involvement in environmental policy. With Takeshita's sudden change, the interest of LDP politicians in the environment also increased with the hope that heightened interest in international involvement and subsequent transnational environmental policy would help to secure a more respected foreign policy profile to become an "international state" (*kokusai kokka*) with global ambitions (Potter 1994: p. 200).

Prime Minister Toshiki *Kaifu* (10 August 1989 - 05 November 1991) also tried to channel the ongoing change of the international system through the decay of the Soviet sphere of influence in Eastern Europe and the dissolution of the Soviet Union towards a more visible role of Japan in the world. Two years after leaving the Japanese government, he headed the Japanese delegation of governmental representatives and businessmen to Beijing for a bilateral environmental cooperation symposium.

The above illustrated that Japan's growing global environmental consciousness was mainly reactive due to massive public campaigns from international environmental NGOs and Japanese companies that had to comply with increasing environmental restrictions around the world. Japanese politicians tried to take advantage of these new challenges and re-gain standing and power even in the international arena. The often referred to value of assisting developing countries to avoid the same mistakes Japan experienced during its rapid economic growth was also laying ground for the transfer of Japanese ET and thereby supporting trade interest and encouraging increased cooperation with countries exporting their resources to Japan securing national industry. Still, the main initiation was from non-state actors and other governments but was often put on the back-burner when Japanese politicians started to closely link domestic and transnational environmental considerations with their own interests.

In the same way Germany's government developed a growing global environmental consciousness, though the pressure from outside Germany was certainly of less importance than in Japan. Moreover, the German public was directly experiencing transnational environmental challenges, which was a driver in the settlement of environmental movements in Germany's political and hence national state system. Nevertheless, the fact that the Germany's and Japan's initiation of transnational environmental consciousness was mainly from non-state actors and supported by environmental catastrophes does not lead to a long-term diminished ability of a state and its machinery to solve life-threatening problems for society. This leads to the questions if and in case how especially Japan responded to these challenges.

10.4 Growing Public Development Assistance

In the first post-war decades, Japan's foreign policy was to a large extent a function of its international trade policy (Kevenhörster 1993a: p. 52). During Prime Minister Yasuhiro Nakasone's term in office (02 November 1982 - 05 November 1987), the predecessor of Takeshita, the aim of a heightened international voice became clear. The increase of expenditure for public developing aid ODA should also support this claim (Nuscheler 1990: p. 37).

For a long period of time, the Federal Republic of Germany's development assistance was dominated by domestic interests, such as the resistance to communism in developing countries in the 1960s. It was also intended to hinder support for a formal acceptance of the GDR in the United Nations (Nohlen 1995: p. 136; Nuscheler 1995: p. 378). And despite the formal launch of a new Federal Ministry of Economic Cooperation (*Bundesministerium für wirtschaftliche Zusammenarbeit (BMZ)*) in 1961, its policy was prescribed by the Federal Ministry of Economics as regarding trade-related issues and the Federal Foreign Office setting the framework of its policy. During his term as Federal Minister for Economic Cooperation Hans-Jürgen *Wischnewski* (01 December 1966 - 02 October 1968) justified the FRG's development cooperation as an appropriate mean for supporting the export and stimulation of economic development. The development of the Basis-of-Relations-Agreement (*Grundlagenvertrag*) between the FRG and GDR, which came into force on 21 June 1973, and the preparations of the UN General Assembly's approval of their UN membership on 18 September 1973 allowed the FRG's development assistance policy to no longer be dominated by inter-German interests. Under Erhard Eppler's term as Federal Minister (16 October 1968 - 08 July 1974) it moved towards a worldwide social and peacekeeping policy, reducing the impact of short-term vested interests; moreover, the role of the BMZ in controlling and assessing Germany's public development assistance was substantially increased (Ibid.: p. 379f). Nevertheless, among others, a lacking support of Germany's business and industry was hindering a re-launch of the FRG's ODA policy.

In the 1970s the economic crisis and the oil-price-shock re-emphasized domestic and economic interests, linking it closely to the trade, economic and foreign policies of the FRG. One started to distinguish between developing countries, tighten up loans for technical and financial cooperation and substantially improved the BMZ's efficiency in implementing the FRG's ODA. Thus, during his term Federal Minister for Economic Cooperation Rainer *Offergeld* succeeded in increasing the FRG's ODA from US$ 2.596 billion (0.36% of GDP) in 1978 to US$ 3.486 billion (0.47% in 1982) or almost 30% (BMZ 2010b; Nationmaster.com 2010). With the takeover of the FRG government by the Christian Democrat / Liberal coalition under German Chancellor Helmut *Kohl* in 1982, the development assistance policy was again mainly to support Germany's international trading interests, also containing communistic movements in the South (Nohlen 1995: p. 137). All in all most analysts agree that the FRG's development assistance was fairly dominated by economic interests in the years 1982-1990, even supporting questionable projects in countries like Zaire and Togo (König 1999: p. 9). With US$ 6.847 billion in 1993 and US$ 6.751 billion, Germany was fourth among the largest donors of DAC Countries in 1994 (MOFA 1994: p. 28; MOFA 1995a: p. 9). In the 1990s the FRG's ODA net disbursements fell from 0.42% in 1990 to 0.28% in 1997 (OECD 1999). In 2008 Germany was second in the DAC list with US$ 12.291 billion just behind the US (MOFA 2010), which equals a portion of 0.38% of GDP.

In comparison with Germany, the growth rates of Japan between 1977 and 1984 were enormous, with Japan doubling ODA distributions up to four times; or it multiplied the ODA budget by nine in two decades.[109] In 1994 in the fourth successive fiscal year, Japan maintained its top place as the worldwide largest donor of developing aid with a budget of US$ 13.47 billion and kept this position until 2000. Based on this one may denote that the ODA budget increased by 14.4% from US$ 11.26 billion in fiscal year 1993 (OECD 1996; p. 2000); however, in 2008 Japan's total ODA net disbursement amounted only US$ 9.58 billion, which equals a proportion of 0.19% of the Japanese GDP and hence half of Germany's ODA proportion in the same year. Both nations are far away from

---

109 Export loans and private transfer (loans of banks and direct investments) are not included in this calculation. From US$ 1.46 billion in 1977 after fulfillment of the reparations to the Philippines in 1976, Japan's ODA budget increased by 800%. It was partly caused by revaluation of the Yen against the US$ as a consequence of the Plaza agreement in September 1985. Nevertheless, the rate of increase is enormous in comparison with other OECD member countries.

the commonly agreed on 0.7% of the Gross National Income (GNI)[110] to support the realization of the Millennium Development Goals (MDGs)[111].

The cuts in Japan's ODA over the previous years result from a) the fiscal crisis forcing the government to cut expenditures not deemed absolutely necessary, b) the Japanese public experiencing aid fatigue as it feels the budget should be spent on services that benefit Japanese society more directly and c) failures in Japan's communication with external stakeholders, internationally, weakening its international position. Despite its past efforts in deepening mutual understanding between Japan and the other DAC countries, Japan feels that a gap exists between the mainstream development thinking and the East Asian development experience, which is widely regarded as a "success story" and to which Japan itself made significant contributions through its ODA. In comparison with Japan, Germany succeeded to stabilize its ODA budget despite the substantial cuts in the overall budget; this was only possible due to special ad hoc funds supporting means against terrorism after 11 September 2001. Moreover, the present budgetary crisis in both Japan and Germany, due to the most recent financial crisis, will not allow substantial increases of the ODA budget in the short-term, making the Millennium Development Goal of 0.7% of GDP unachievable in both the short and mid-term, which was already postulated at the UNCED in Rio de Janeiro in 1992.

10.4.1 Quantification of Official Development Assistance (ODA)

Japan and Germany's aid performance includes public developing assistance on both bilateral and multilateral levels. Bilaterally, ODA is divided into grants (*joseikin*) and loans (*rōn*) in Japan, whereas one distinguishes between financial cooperation (*Finanzielle Zusammenarbeit = FZ*) and technical cooperation (*Technische Zusammenarbeit = TZ*) in Germany. Hence, on the surface the organization of Germany's ODA appears more functional than financial in comparison to Japan, though this needs to be proven.

---

110 GNI equals Gross Domestic Income (GDI) or GDP plus income receipts from the rest of the world minus income payments to the rest of the world.
111 In September 2000, building upon a decade of major United Nations conferences and summits, world leaders came together at United Nations Headquarters in New York to adopt the United Nations Millennium Declaration, committing their nations to a new global partnership to reduce extreme poverty and setting out a series of time-bound targets - with a deadline of 2015 - that have become known as the Millennium Development Goals. See for example http://www.un.org/millenniumgoals/bkgd.shtml – 02 August 2010.

The Japanese grants include technical cooperation and financial resources, which the developing countries do not have to repay (MOFA 1995b: p. 13). In comparison with other DAC member countries, Japan had many actors in its ODA administration in the 1990s and this still remains the case. There were and still are four ministries – the Ministry of Foreign Affairs (MOFA), the Ministry of Finance (MOF), the Ministry of International Trade and Industry (MITI) and the Economic Planning Agency; in total approximately 15 ministries and agencies as well as several implementing organizations such as the Japan International Cooperation Agency (JICA) and the Overseas Economic Cooperation Fund (OECF) were involved in the development assistance programme in the 1990s (OECD 2000: p. 29ff) and are still active today. In the 1990s over 50% of tasks related to the technical cooperation were implemented through JICA (JICA 1994: p. 17).

Projects in e.g. public health teaching and research, agriculture and the protection of the environment were regarded as pillars of Japanese development aid in the 1990s. The six classes of Grant Aid (developing aid through donations) were:

1. General Grant Aid,
2. Support of the fishery nature,
3. Help after disasters,
4. Support of cultural activities,
5. Help with food,
6. Improvement of food production (JICA 1995a: p. 3).

Today Japan's Grant Aid is divided into the following ten categories:

1. General Grant Aid Projects,
2. Non-Project Grant Aid,
3. Grant Aid for Grassroots Human Security,
4. Grant Assistance for Japanese NGOs,
5. Grant Aid for Human Resource Development,
6. Grant Aid for Cooperation on Counter-Terrorism and Security Enhancement,
7. Grant Aid for Disaster Prevention and Post-Disaster-Reconstruction,
8. Grant Aid for Community Empowerment,
9. Grant Aid for Fisheries,
10. Cultural Grant Aid (JICA 2010: p. 24).

Though the number of specific areas has increased since the 1990s, food-related activities are no longer explicitly mentioned. Remarkable is the additional provision of Grant Aid to NGOs and communities, obviously reflecting a substantially increased role of the two in Japan's transnational development cooperation. This points to the assumption that they gained power, possibly also in comparison with the government. Nevertheless, these changes of key categories do not necessarily provide evidence for a decreased role of the nation-state in this policy field.

In contrast to Japan Germany's bilateral ODA has had only one main actor since the 1990s – the Federal Ministry for Economic Cooperation and Development (BMZ) at the policymaking level which has a stronger position than the aid administration in many DAC countries. Nevertheless, for example, the German Foreign Office requests a close consultation with the BMZ on aspects of foreign policies; similar to Japan Germany also has multiple systems for its ODA implementation. Still today, there is an institutional separation between financial and technical cooperation, with the latter being implemented by a large number of institutions.[112]

Poverty reduction, gender and the protection of the environment are crosscutting tasks permeating all German aid activities. Direct development cooperation between Germany and its partner countries consists of two elements: Technical Cooperation[113] and Financial Cooperation[114].

---

112 Financial assistance is exclusively administered by the Kreditanstalt für Wiederaufbau (Bank for Reconstruction). The main channel for technical assistance was still until recently the Gesellschaft für Technische Zusammenarbeit (German Technical Cooperation (GTZ)), but important roles were also played by the German Foundation for International Development (DSE), the German Development Service (DED), the Centre for International Migration and Development (CIM), the German Investment and Development Corporation (DEG), the five Political Foundations, the Carl-Duisberg Society (CDG), the German Academic Exchange Service (DAAD), Protestant and Catholic organizations and other NGOs partly until today. Last but not least, the KfW also finances project-related technical assistance to its counterparts.

113 German Technical Cooperation has the task of developing the capacities of individuals, organizations and societies in partner countries. They are to be enabled to improve their own living conditions and to realize their own objectives by making efficient and sustainable use of resources. Through Technical Cooperation, Germany transfers technical, economic and organizational knowledge and skills. Since the UNCED it aims at taking into particular account the need to involve civil society and to empower women in partner countries. (Ibid.). The government-owned German Agency for Technical Cooperation (Deutsche Gesellschaft für Technische Zusammenarbeit GmbH (GTZ)) was then generally contracted to implement projects pertaining to these cooperation measures agreed at government level. Activities include

Expert advisory services,

Financial support of qualified local executing agencies,

Institution building,

As illustrated in Table 8 in Japan the so-called bilateral Grant Aid amounted to US$ 5.423 billion in 1994, US$ 4.949 billion in 1998 and US$ 7.764 billion in 2008 (MOFA 1995a: p. 108; MOFA 1999: p. 124; OECD 2010: p. 195). Germany's bilateral grants amounted US$ 3.549 billion in 1994, US$ 3.315 in 1998 and US$ 9.932 in 2008 (OECD 1998: p. A28; OECD 2002: p. 77; OECD 2010: p. 194).

Table 8: Bilateral Grant Aid – Japan & Germany (1994, 1998, 2008)

Country	1994	1998	2008
	(in bUS$)	(in bUS$)	(in bUS$)
Japan	5.423	4.949	7.764
Germany	3.549	3.315	9.932

(Composed out of data of MOFA (1995a); OECD (1998), (2002) & (2010))

Japanese and German Technical Cooperation (TC) projects are created with partner countries taking into account local conditions, utilizing the knowledge, experience and technology of both Japan and Germany and the developing countries toward the resolution of issues within a specified time frame. An important goal of Japan and Germany's TC is to impart know-how and national technology with persons playing a lead role in the area of "technical cooperation" within the developing countries. These so-called "counter-partners" or "experts" represent the interface for the dissemination of technology there and should thus also support national development.

Capacity development,
Provision of equipment and material,
Inputs through studies and reports (Deutscher Bundestag 1995: p. 52ff).
Projects and programmes focus on those fields and those regions that have been identified as priority areas of bilateral development cooperations with the partner country. Even until December 2010 this approach has not substantially changed since1990, though the GTZ was merged with the DED and Inwent on 1 January 2011 now forming the German Society for International Cooperation (GIZ).
114 German Financial Cooperation is an instrument involving state bodies only. It can be used, for instance, to finance investments, particularly in infrastructure and financial systems, to finance materials and equipment or to establish effective structures. The volume of funds is agreed and laid out in a contract between Germany and the partner country. The funding usually takes the form of soft loans; the poorest developing countries (LDCs) are also granted funding in the form of a non-repayable grant. One special form of Financial Cooperation is called "programme-oriented joint financing". Here the funding is not made available for individual projects but is paid into the partner country's budget (BMZ 2010a).

Additionally, Japanese and German experts and volunteers visit the developing countries in order to support the planning and preparation of developing programmes. The Japanese MOFA defines technical cooperation as a "*cooperation of technology and the human heart*" (MOFA 1995: p. 132) because of its contribution to the understanding between people.[115]

In addition to this dispatch of experts and the provision of equipment, the Japanese technical cooperation also focuses on the acceptance of training participants, whereas this only plays a marginal role in Germany's TC. Since the adjustment of Japan's ODA Charta[116] in 2003, the partnerships with NGOs and the coverage of local activity expenses are also among the major input areas of Japan's TC (JICA 2010: p. 128).

As illustrated in Table 9, the share of TC of Japan's entire bilateral ODA amounted to 31.2% or US$ 2.985 billion in fiscal year 1994; in 1998 US$ 2.781 billion (32.3% of bilateral ODA) and in 2008 US$ 2.986 billion (44.96%). In 1994 the share of TC of Germany's entire bilateral ODA amounted to 59.9% or US$ 2.126 billion, in 1998 US$ 1.988 billion (60% of bilateral ODA) and in 2008 US$ 4.187 billion (44.6%) (OECD 1998: p. A28; OECD 2002: p. 77; OECD 2010: p. 194).

Table 9: Bilateral Technical Cooperation – Japan & Germany (1994, 1998, 2008)

Country	1994		1998		2008	
	(in bUS$)	(in % of bilateral ODA)	(in bUS$)	(in % of bilateral ODA)	(in bUS$)	(in % of bilateral ODA)
Japan	2.985	31.2	2.781	32.3	2.986	44.9
Germany	2.126	59.9	1.988	60.0	4.187	44.6

(Composed out of data of OECD (1998), (2002) & (2010))

ODA loans (frequently called credits, loans or subsidies) made available for developing countries are financial resources on a loan basis, which are expected to be re-paid over a long period at low interest rates. Thereby, projects are the foreground for the improvement and development of the economic and social infrastructure, which includes the building of streets, dams, arrangements for

---

115 Trainees at ICETT have to spend several months in almost total seclusion in the mountains around Yokkaichi. Spontaneous purchasing and trips are impossible due to the distance and lack of bus lines. Thus, contact with Japanese outside the Centre is rather difficult, which is also criticized by ICETT staff.

116 Japan's ODA Charter, approved by the Cabinet in 1992, has been the foundation of Japan's aid policy for more than 10 years. For more details see http://www.mofa.go.jp/policy/oda/white/ 2009/html/index_shiryo.html – 05 August 2010.

telecommunication, the improvement of agriculture etc. Recently, loans are used increasingly for financing the purchase of products to compensate the imbalance of the international balance of payments and thus to equalize it. Japan's budget for bilateral loans was under the jurisdiction of the Ministry of Finance; operated by the Overseas Economic Cooperation Fund (OECF) after consultation with the Ministries of Foreign Affairs, Finance and International Trade and Industry as well as the Economic Planning Agency until October 1999. As part of Japan's organizational restructuring of its ODA, the OECF and the Japan Export Import Bank (JEXIM) were replaced by the Japan Bank for International Cooperation (JBIC) (*Kokusai kyōryoku ginkō)*; since 2008 the international wing of Japan's Finance Corporation was administered by the Ministry of Finance.

The total amount of bilateral development aid amounted to US$ 9.680 billion in Japan and US$ 4.144 billion in Germany in 1994[117] and totalled 71.87% of the entire ODA budget in Japan and 60.8% in Germany; in 1998 in Japan US$ 8.553 billion or 70.4% and US$ 4.544 billion or 68.4% in Germany (MOFA 2000: p. 124; OECD 2002: p. 77) and in 2008 US$ 6.641 billion or 70.9% in Japan (JICA 2010: p. 25) and US$ 9.063 billion or 64.8% in Germany (OECD 2010: p. 194) – see Table 10.

Table 10: Total Bilateral Development Aid – Japan & Germany (1994, 1998, 2008)

Country	1994		1998		2008	
	(in bUS$)	(in % of ODA)	(in bUS$)	(in % of ODA)	(in bUS$)	(in % of ODA)
Japan	9.680	71.9	8.553	70.4	6.641	70.9
Germany	4.144	60.8	4.544	68.4	9.063	64.8

(Composed out of data of JICA (2010); MOFA (2000); OECD (2002) & (2010))

Indonesia, with US$ 886.17 million, was in the third place behind China and India on the list of the main recipients of Japan's bilateral ODA in 1994 and was among the 34 states for which Japan is "Top Donor" (MOFA 1996: p. 10). Indonesia was Germany's top ODA recipient in the mid-1990s with US$ 0.413 billion, and second after China with US$ 0.226 billion in 1998 (OECD 2001: p. 77). In Japan in 1998 Indonesia gained the second place spot with US$ 828.47 million after China but before Thailand (MOFA 2000: p. 127); in 2008 this list of top recipients of Japan's Grant Aid was led by Iraq, Bangladesh and Afghanistan with Indonesia no longer among the top 30 recipients (MOFA 2010). Iraq was also the top recipient of Germany's ODA in 2008 with Cameroon and Chi-

---

117 Without Eastern Europe. Including Eastern Europe the total amount is US$ 9.680 billion. (MOFA 1995b: p. 12).

na following; however, differing from Japan, Indonesia was still ranked number six with US$ 0.246 billion (OECD 2010: p. 112).

A further element of development assistance are membership fees and donations from international organizations including the United Nations Environment Program (UNEP), the United Nations Development Program (UNDP), the United Nations Educational, Scientific, and Cultural Organization (UNESCO), the World Health Organization (WHO) and further United Nations (UN) Organizations, the World Bank, the International Development Assistance (IDA) and the Asian Development Bank (ADB). At US$ 3.681 billion, the share of this total multilateral developing aid is 27.8% of ODA expenditure in 1994; in 1998 US$ 2.567 billion or 28.9% and in 2008 US$ 2.724 billion or 28.7% (MOFA 2000: p. 121; JICA 2010: p. 25). In 1994 Germany contributed US$ 2.674 billion or 39.2% of its ODA budget to multilateral institutions; in 1998 US$ 2.100 billion or 31.6%, and in 2008 US$ 4.918 billion or 35.2% (OECD 1998: p. A28; OECD 2001: p. 77; OECD 2010: p. 194).

Table 11: Multilateral Development Aid – Japan & Germany (1994, 1998, 2008)

Country	1994		1998		2008	
	(in bUS$)	(in % of ODA)	(in bUS$)	(in % of ODA)	(in bUS$)	(in % of ODA)
Japan	3.681	27.8	2.567	28.9	2.724	28.7
Germany	2.674	39.2	2.100	31.6	4.918	35.2

(Composed out of data of JICA (2010); MOFA (2000); OECD (1998), (2001) & (2010))

Germany's strong support of multilateral institutions must also be seen as its support of the ODA channels through the European Community, through which Germany contributed substantially to the funding of Central and Eastern European Countries and the Newly Independent States in the 1990s. Despite substantial cuts of Japan's ODA since the late 1990s, the level of contribution to multilateral institutions has not substantially decreased. In consequence Japan, but also Germany, see an increased role of multilaterals in a transnationalizing world, making them also a key recipient of their respective ODA. In how far these adjusted roles of multilaterals are impacting those of Japan as a nation-state will be analyzed in the following.

The above illustrated not only the organization of Japan's ODA but also substantial changes in the ODA budget since the 1990s. These changes suggest foremost that an adjusted prioritization downgraded ODA on Japan's political agenda, at least in budgetary terms. This observation is also supported by lacking concepts how to finally reach the agreed-upon MDG target of 0.7% of GDP.

On the other hand Japan's officials have hardly any other chance than substantially cutting the ODA being confronted with public debts of 189.3% of its GDP in 2009 making Japan worldwide the second country after Zimbabwe with the largest public debt in comparison to GDP. In consequence there are increasing calls from outside Japan for strict budgetary and fiscal measures. Germany showed that the substantial decrease of its ODA budget was inevitable due to the economic and financial crisis, though Germany's public debt rate of 72.1% is the highest ever, yet far less than Japan's, still allowing to stabilize its ODA until recently (CIA 2010b). Nevertheless, it would be misleading to directly draw conclusions about the quality of development assistance and even e.g. transnational environmental and trade policies out of a quantification based on official statistics. The quantification indicates a certain political trend in Japan of a reduced political consideration of ODA at the face of enormous national debts, but it does not necessarily reflect the political efforts and its outcomes (Kevenhörster 2009: p. 41), necessary for the verification or falsification of Dehnhard's theory. The continuation of the policy-cycle analysis including its analysis of actors involved in the transfer of environmental technology will therefore support such.

10.4.1.1 Growing Environmental Orientation in ODA

With the adoption of the Council Recommendation on Environmental Assessment of Development Assistance Projects and Programs and the Council Recommendation on Measures Required to Facilitate the Environment Assessment through the OECD in the years 1985 and 1986, OECD members were called to take environmental aspects increasingly into their development aid programmes (OECD 1985). This happened during Japan's second phase of rather environmental ignorance and hence came unprepared to the state's institutions. Before 1988 the Japanese public paid little attention to transnational environment problems as illustrated in Chapter 10.3.

As a reaction to this OECD demand, JICA authorized one "Study Group" to demonstrate ways of linking environmental themes in developing cooperation. The concluding report of this investigating committee identified and discussed possibilities a) to improve the (cap)abilities of developing countries, b) to cope with environmental problems, c) to expand and improve environmental projects, d) to use EST efficiently and e) to push the protection of natural resources (JICA 1988).

At the same time a discussion round evolved through the Environment Agency under the title *Basic Directions for Environmental Considerations in Development Assistance* with the strengthened linking of environment problems to developing cooperation and the transfer of appropriate technologies to NICs (Forrest 1991: p. 30).

Japan then undertook the task to set up future development aid decisively on its own experience with pollution. In 1988 global environment aspects were also for the first time included in a White Paper from its Environment Agency. Japan's successive commission to the WCED, called Committee on Global Environment Problems, recommended an expansion of Japan's eco-political development assistance in a report published in June 1988. However, this "ad hoc" commission submitted just as little as before, namely short-term created teams' concrete suggestions to the acceptance of global environment problems on the agenda for development aid. However, it encouraged the 16 ministries and state authorities[118] who were jointly responsible for the formation of the development aid policies (EA 1988: p. 43f.) to determine environmental guidelines. Simultaneously, the applications of potential recipient countries were examined on environment compatibility.

According to the order of the respect before the sovereignty of a developing country, applications must be submitted to the respective Japanese embassy to obtain the benefit of the Japanese development aid fund.

In comparison with Japan, sustainability was already playing a role in Germany's ODA before the release of the "Brundtland Report" (1987) and the UNCED (1992). The time-limited projects of Germany's development cooperation had to ensure sustainable effects and hence reverse the consequences that e.g. dwells and school buildings were not maintained over a long period of time. Bad investments led to the requirement that Germany's development assistance projects had to ensure positive impacts in the long-term (Schubert 1984). This requires the examination of whether follow-up costs are fundable, whether the local personnel has developed sufficient capacity to continue activities, whether the recipients of the services provided are capable to make the necessary contributions and whether the implemented procedures and consumed resources allow a regeneration of the ecosystem. This was and still is a rather demanding claim of Germany's ODA, especially because many involved in development assistance projects are impatiently waiting for action and do not feel in the position to undergo a lengthy impact assessment. However, similar to Japan, an explicit assessment of the environmental impacts of Germany's ODA, in addition to eco-

---

118 The so called Yonshocho (four ministries) has a coordinating and leading function. This is a committee, meeting regularly and consisting of vice-ministers and ministry officials from MITI, MOFA, Economic Planning Agency (EPA) and the Ministry of Finance (MOF). It is responsible for applications to finance projects of development aid. The allocation of grants and TC is incumbent on MOFA exclusively.

nomic and social effects, has had to be undertaken only since the worldwide introduction of sustainable development as the new paradigm (Rauch 2009: p. 76). Since then the number of resource protection projects has increased. On the other hand the deregulation of the International Monetary Fund, the World Bank, the World Trade Organization supported by the national governments introduced a short-time valorization of natural resources, which stands in sharp contrast to this aforementioned paradigm.

10.5 Political Motives of Initiation

In the age of multi-polarity after the breakdown of the former Soviet sphere of influence, Japan, as foreign trade, technological and ODA world power in the 1990s, was also confronted with more global responsibility (Kevenhörster 1993a: p. 15). One reason to put the environmental problems on Japan's political agenda was because some Japanese politicians wanted to give their foreign policy a new profile. Environmental policy was thought to weaken the reproach of a timid and reactive foreign policy (Maddock 1994: p. 37); with the growing global interest in environmental problems, there was strong involvement in this policy

Likewise, a foreign policy motive might be the extenuation of the deep-seated fear of Japan's neighbours of a larger international responsibility of Japan through the contribution of growing involvement of Japanese politicians in the global environmental policy (Yamada 1995a: p. 12). With the help of a decisive weighting of eco-orientated aspects in Japan's development aid, the criticisms of the quality of Japanese ODA, despite having the largest worldwide financial ODA budget, became silent; it should be obvious that fulfilment of development goals had to demonstrate the growing effectiveness.

Security aspects have also contributed to identifying the need for action in transnational environmental policy. Close regional cooperation was expected for an enduring stabilization in the region, especially because of certain trouble spots in East and Southeast Asia; such trouble spots also included the development on the Korean peninsula as well as the tensions between the People's Republic of China and Taiwan in the Pacific region (Nashima 1995: p. 6). An ecologically compatible growth should have contributed to sustainable development in such rapid growing NICs like Indonesia and thus, should have weakened possible conflict potential through e.g. environmentally-induced migration (Fasbender 1995: p. 361ff). Additionally, a more sustainable use of natural resources, which Japan imports as a country deficient in natural resources, was expected to contribute to a political, economic and ecological stabilization of this region; in fact, Prime Minister Nakasone called this an existential mark of Japan's foreign policy. Thereby, Japan was considered to open and secure energy and raw material sources.

Since the mid-1980s the tremendous growth market for environmental technology has surely also pushed the interest of more Japanese politicians towards involvement in environmental policy, as it did with Germans, and must be considered as a foreign trade motive in this investigation. Besides, Japan and Germany's bitter experiences in the 1950s/1960s did not play a neglecting role, because such a purely economically-oriented development in countries like Indonesia and China could mean widespread ecological collapse.

In domestic policies, numerous Japanese politicians have tried to polish their negative image influenced by corruption through activities in the global environmental policy arena and thereby simultaneously weaken the argument of domestic immobilism through the formulation of a transnational policy. Out of today's view they failed with this attempt, primarily because Japanese politicians are not recognized for their leading roles in environmental policies. This might also be connected to the constant changes of Japanese governments, which may hinder the development of an international profile necessary for a leading role.[119]

An explanation of Japan's transnational environmental policy through the transfer of environmental technologies to NICs with the focus on foreign trade motives would only be considering the complexity of the occasions and their interdependence too short sightedly. In the 1990s Japan was no longer following the primacy of foreign trade interests in its development cooperation.

Altogether, motives from six different fields have contributed to putting environmental technology transfer onto the political agenda in Japan. These were 1) foreign trade politics, 2) development politics, 3) domestic policy, 4) security policy, 5) environmental policy and 6) economic policy. Motives can be defined as an incitement of action towards a desired goal; values are contributing to the development of this desired goal. In consequence trade, assistance, security and environmental values were important for initiating political action towards environmental technology transfer to NICs. Beyond these, the sharpening of the foreign profile of a nation through contested politicians trying to re-gain power was another driver.

In contrast to Japan, Germany's political motives were quite inspired by environmental considerations together with aspects of development assistance.

---

119 Japan has had 15 Premier Ministers only within the present Heisei-Period, the current era name. The Heisei era began on 8 January 1989, the first day after the death of the then reigning Emperor Hirohito.

However, the trading potentials together with the request to play a more important international role, also reflecting its economic power, plus the securing of votes and hence support by the majority of voters in a phase of upheaval in the landscape of German political parties, were also important motives. Nevertheless, the comparison between Japan and Germany points out one great difference: Until only recently Germany was not very active in securing natural resources for its industrial production chains, whereas it is high on Japan's political agenda for decades. Initiatives towards a Zero Emissions and Low Carbon Society together with the 3R[120] work are part of Japan's long-term strategy as a resource poor country.

The analysis of the initiation of Japan's political action towards the transfer of environmental technology to NICs illustrated the important role of non-state actors and influence from outside Japan. Without a doubt these contributed to the initiation of the political process. However it would be hasty considering this as evidence for the diminishing of the state's role in this transnational sector without analyzing the policy formulation and the resulting implementation.

---

120 The 3R Initiative aims to promote the "3Rs" (reduce, re-use and recycle) globally so as to build a sound material-cycle society through the effective use of resources and materials. It was agreed upon at the G8 Sea Island Summit in June 2004 as a new G8 initiative. It was formally launched during a ministerial meeting in Japan in the spring of 2005 and since then Japan is key driver towards this aim.

11 Estimation

Only at the end of the 1980s, political protagonists in Japan recognized the transnational environmental problems as relevant for action. The second step of the Policy-Cycle Model – the estimation – analyzes the situation in order to identify possible alternatives for action. The estimation allows obtaining a deeper insight into the values and powers of the identified actors.

11.1 International Conferences

An elementary forum to discuss alternatives for action within transnational environmental policies is represented by international conferences. Thereby the embedding of national policies in an internationally harmonized approach shall respond to the transnational nature of the problem (Kaiser 1991: p. 13). Simultaneously, these scientific, political, industrial/business or even mixed conferences offer important fora to present national initiatives to a general public and thus, to strengthen their reputation in foreign policy and to gain profile for domestic interests.

Table 12: Japan's Environmental Cooperation (1989-1998)

Fiscal Year	Grants JP¥ 100 Million	Loans JP¥ 100 Million	Technical Assistance JP¥ 100 Million	Multilateral Assistance JP¥ 100 Million	Total JP¥ 100 Million	Accumulated Total JP¥ 100 Million
1989	227.4	920.2	100.3	46.0	1,294	1,294
1990	228.4	1,243.8	132.4	49.2	1,654	2,948
1991	241.5	666.6	141.2	77.5	1,127	4,075
1992	310.6	2,212.5	174.1	105.7	2,803	6,878
1993	377.1	1,526.5	214.1	162.0	2,280	9,158
1994	414.3	1,055/1,054*	234.2/218.7	253.3	1,941/1,958*	11,099/11,116*
1995	428.2	1,708.2	222.9	400.3	2,760	13,859/13,876*
1996	360.7	3,864.7	253.4	153.8	4,632	18,491/18,508*
1997	364.6	1,623.4	300.7	158.1	2,447	20,938/20,955*
1998	289.9	3,280.9	304.2	263.1	4,138	25,076/25,093*

*MOFA provides different numbers in its Annual Report for the year 1994

(Composed out of data of MOFA 1999: p. 69; 1996: 189; 1995: p. 196)

During the Arche summit in Paris in July 1989, the Japanese government announced through Prime Minister Sôsuke Uno (03 June 1989 – 08 August

1989) the provision of JP¥ 300 billion (US$ 2.2 billion) for eco-political ODA within the time period 1989-1991 (MOFA 1990: p. 44).

Table 12 shows that Japan overstepped, however, this target by more than 30% with an accumulated total of JP¥ 407.5 billion (approx. US$ 3 billion) of total pledges for bilateral and multilateral environmental protection projects in its developing cooperation investments at the end of 1991. Nevertheless, this quantitative increase in Japan's eco-political bilateral and multilateral ODA could not mislead, however, with regard to the countless qualitative defects in the 1990s.

Upon entering the 1990s plans of Japanese development assistance still prevailed, which appeared ecologically questionable and were expected to negatively affect the social situation of the regions and their population in the long-term. A few examples include the building of gigantic lakes in arid zones of Africa, the dam construction at the Yangtze River in China, gigantic infrastructure projects connecting Indonesian islands through bridges and afforestation projects adversely impacting exotic trees in e.g. India and Thailand.

Considering these evident defects of quality, the proclamation of a year of "environmental diplomacy" by the Japanese Ministry of Foreign Affairs in January 1989 appeared as bare lip service. Concrete attempts towards a transnational environmental policy through eco-political development aid were lacking. Questions concerning Japan's ODA were forwarded to the speech writing departments of the Japanese government instead of directly to the responsible ministerial officials. In the eyes of Richard *Forrest*, this shows that Japan's policies were hardly developed above pure rhetorical announcements (Forrest 1991: p. 32).

The Conference on the Global Environment and Human Response Toward Sustainable Development (11-13 September 1989), arranged by Prime Minister Takeshita, demonstrated no alternatives but elucidated his particular interest in recovering political profile through his involvement in global environmental policy. The seriousness of this event organized by Japan's governing party (LDP) to discuss possible activities in developing cooperation must also be questioned because of the exclusion of non-governmental organizations (NGOs) (Holiman 1990: p. 288). It also illustrates the perception of the then-Japanese leading political group that environmental issues, especially those of transnational character, fall within the exclusive domain of the government.

At other global summit meetings Japanese officials did not refrain from testifying the will of their country to take on a leading role in global environmental protection. But at the sight of the Gulf War[121] and the rapid changes in Eastern Europe, these were, however, announcements more of a subordinated nature failing to attract the intended large attention. Moreover, concrete changes in policies and measures, which identified global environmental protection as an urgent task, were not additionally obvious and supportive of these announcements.

This changed with the G7 summit in July 1991 in London. During the summit the Japanese delegation announced its intention to strengthen Japan's efforts on cooperation in development assistance with the Third World countries and to thereby benefit from Japan's technologies and experiences, which Japan developed through the handling its environment problems in the years of quick economic growth between 1950 and 1970. Additionally, Japan was seen to be in the position to react variably on the various developing steps of the recipients and to devote itself especially to poverty and population growth issues, which have to be seen as directly related to environmental problems. During this summit Japan granted priority in its development aid e.g. to the protection of forests and reforestation, the preservation of natural energy sources, the employment of clean technology, pollution control and soil protection (JICA 1995b: p. 2). Hence, environmental technologies were a considerable focus.

During the Rio-Conference in 1992, Tokyo attracted attention by announcing it would provide approximately JP¥ 900 billion to JP¥ 1 trillion (US$ 6.75 billion - US$ 8.5 billion) for eco-political ODA until 1996. With JP¥ 978 billion at the end of fiscal year 1995 this goal - the five-year plan - was already crossed (see Table 12). Furthermore, the Japanese delegation underlined the importance of the transfer of technology for the protection of the natural environment (Pollack 1992: p. 11). As a first concrete measure, the Japanese government introduced the establishment of an International Environmental Technology Centre (IETC) under UNEP's umbrella as an interface between the producers and users of environmental technology. With the acceptance of the final declaration, called Agenda 21 or Rio Declaration for Sustainable Development, the Japanese government obliged itself to set up a new, global partnership, to adopt new methods for cooperation between the nations, crucial questions of society and people and thus, to follow the integral and mutually dependent character of life on earth (Strong 1992: 234ff.). Besides a preamble, the Agenda 21 includes 27 principles altogether, which were expected to serve as the orientation of the international community at the threshold to the 21st century (UNCED 1992). The assembled

---

121 In response to the invasion of Kuwait by Iraqi troops, a UN-authorized coalition force from 34 nations led by the United States and the United Kingdom expelled the Iraqi troops from the Kuwait territory and advanced into the Iraq. This is usually referred to as Persian Gulf War (2 August 1990 – 28 February 1991) or simply the Gulf War.

leaders also signed the Convention on Climate Change and the Convention on Biological Diversity and endorsed the Forest Principles[122].

In Rio de Janeiro it was agreed that a five-year review of the Earth Summit progress would be made in 1997 by the United Nations General Assembly meeting in special session. This special session of the UN General Assembly, the so-called Earth Summit+5 in 1997 took stock of how well countries, international organizations and sectors of civil society have responded to the challenge of the Earth Summit. When this Earth Summit+5 took place, Japan exceeded the targeted amount of ODA in the environmental field by more than 40%, providing JP¥ 1.44 trillion (app. US$13.3 billion). In his statement Japanese Prime Minister Ryûtarô *Hashimoto* stressed the necessity to develop innovative environmental technologies and promote their transfer to developing countries in order to foster sustainable development. The Japanese delegation also promoted a new plan entitled *Initiatives for Sustainable Development Toward the Twenty-first Century (ISD)*, further promoting the transfer of environmental technologies for the prevention of pollution and technologies related to the conservation of energy and new energy sources. Hashimoto referred to the budgetary constraints his government started to face in the mid-1990s, but nevertheless pledged to give greatest possible consideration to ODA in the area of environment and announced Japan would make it a national policy to cooperate in the promotion of sustainable development (Hashimoto 1997). When he was explaining that Japan is prepared to present not only its successes but also its failures, and to cooperate with others so that its mistakes will not be repeated, he was referring more to Japan's history of environmental pollution and not its late launch of transnational environmental policies and shortcomings of such since.

During the WSSD in Johannesburg in 2002, the Japanese delegation introduced the Koizumi Initiative of concrete action to be taken for Sustainable Development and towards global sharing. Though contrary to earlier statements of other Japanese Prime Ministers, these plans did not directly refer to environmental technology transfer; rather they emphasized the sharing of experiences on pollution and its recovery from as well as its successful experience of cooperation with other Asian and African countries as well as other regions (MOFA 2002).

In 1998 in its report to the German Parliament, the government described the transfer of environmental technology as one focus area of bilateral environ-

---

[122] Guiding objective of these principles is to contribute to the management, conservation and sustainable development of forests and to provide for their multiple and complementary functions and uses. For more details see http://www.un.org/documents/ ga/confl51/aconfl5126-3annex3.htm – 10 July 2010.

mental cooperation (Deutscher Bundestag 1998a: p. 207). This report and also the report on the government's development cooperation policy each devotes one sub-chapter to NGOs, illustrating them as key-actors in Germany's transnational environmental policy through their grassroot work (Deutscher Bundestag 1995: p. 93ff).

In 2002 at the WSSD, Chancellor Gerhard *Schröder* (1998-2005) focused on energy announcing increased cooperation in the energy field, calling it a strategic partnership. In Johannesburg Schröder made a pledge of EUR 500 million to promote cooperation on renewable energies but did not refer to any specific transfer of technologies (Schröder 2002).

The above illustrates that in contrast to Japan, Germany did not stress the transfer of environmental technology as a key-element of its transnational environmental policy. Also monetary aspects of its ODA did not play such a prominent role as in the announcements of the Japanese Prime Ministers. Though the Chancellors stressed Germany's global responsibility, they did not explicitly refer to Germany's experiences with wide-spread pollution and its subsequent coping as a model for NICs and developing countries.

The amounts of environmental ODA in the 1990s and high-level political statements on environmental ODA can be regarded as evidences for a reorientation of Japan's environmental policy on the transnational level. Transnational environmental problems were for the first time regarded as serious challenges from Japan's political class combined with concrete plans for action in which the transfer of environmental technology plaid a key-role. The political statements also emphasized the importance of transnational cooperation and globally harmonized action towards sustainable development, though this has become part of each political statement since Rio and hence almost a platitude at the absence of concrete measures. And it does not come to large surprise that Hashimoto and others also made clear that one must be in the position to afford the necessary measures, latently emphasizing that it will be Japan's transnational environmental action which will suffer as a result of necessary measures to revitalize its economy. This way the value of transnational environmental action is regarded as less important than domestic prosperity. The numbers and statements do not provide any insights into the quality of the final implementation and the necessary frameworks in which they were developed. Germany especially stressed the energy sector for its activities, whereas Japan kept this open. Both, Germany and Japan did emphasize the role of non-governmental actors such as NGOs or players in the private sector for transnational environmental policy and prominently referenced international organizations such as the Global Environment Facility[123], the World Bank, UNEP, UNDP etc.

11.2 Institutional Changes in Japan's ODA

Prime Minister Uno's assurance to increase the share of eco-political development assistance in Japan's ODA-budget was followed by changes in Japanese institutions in August 1989. Within the JICA Planning Department, the Environment Unit, as a subdivision for the area of environment, was established and for each JICA section a person responsible for environmental affairs was determined (JICA 1995b: p. 4). In May 1991 this subdivision also became responsible for women in development (WID), poverty alleviation and further unnamed global aspects. These changes were also followed in 1993 by the establishment of the section for "Environment, WID and further global aspects". These first institutional changes were introduced by the *Sectional Study for Development Assistance on the Environment*. Since 1989 this study assessed JICA's organizational structure and examined its ecological activities. Simultaneously, JICA dispatched the *Project Formulation Survey Teams* to introduce environmental aspects into the concrete project work. Likewise, JICA established guidelines for developing studies together, which embedded the environment problem into the project cycle very early (JICA 1994a: p. 25); in 1993 the fishery sector appeared as the twentieth area on the list. From 1989 onwards an environmental impact assessment had to be preceded for the allocation of bilateral credits for development projects financed through the OECF. Besides this finance institute, whose annual budget of US$ 4.67 billion was 35.3% of the entire Japanese ODA budget in 1994 (OECF 1995: p. 25), almost half the budget of the World Bank, published sixteen environmental guidelines. These were referring to a series of aspects, in which the implementation of development projects could influence the environment. It was foreseen that the governments of developing countries should diligently take environmental aspects into account in the phases of project development and project preparation; in this respect the environmental instructions of the applicant and the instructions of international conventions were to be observed. However, environmental specialists were lacking, who were in-

---

123 The Global Environment Facility (GEF) is an independent financial organization providing grants to NICs and developing countries for projects related to biodiversity, climate change, international waters, land degradation, the ozone layer and persistent organic pollutants (POPs). By this GEF links local, national, and transnational, partly global, environmental challenges and promotes sustainable livelihoods. For more details see http://www.thegef.org – 06 September 2010.

creasingly bound through institutional changes in the project work; these specialists should have been responsible for assessing environmental influences of different development projects (Murdo 1994: p. 6). In 1990 the OECF appointed an Advisory Committee on Environmental Issues to receive and compile advice from experts of completely different fields in environmental tasks. However, it was only in October 1993 that the OECF created a Section for Environment and Social Development to devote itself to environmental problems and WID; in August 1995 the environment guidelines were revised.

Japan played a key-role in the preparation of a New Development Strategy in the OECD's Development Assistance Committee (DAC) entitled "Shaping the 21St Century: The Contribution of Development Cooperation" by proposing specific developments goals. This was also a response to the growing public fatigue on development cooperation especially because of doubts on its effectiveness as e.g. African development did not have many success stories in the 1990s. This new DAC Strategy acknowledged that primary responsibility for development rests with the recipient countries; nevertheless the close cooperation between donors and recipients in sharing responsibility was seen a mandatory for working hand-in -hand to achieve the desired development. This Strategy proposes concrete numerical development goals for the industrialized and developing countries to share, including poverty reduction, prevalence of primary and secondary education, lowering of the infant and maternal mortality rates and planning national strategies for sustainable development in order to ensure that loss of forest, water and other resources are reversed by 2015 (MOFA 1996: p. 17).

In January 1998 an ODA Reform Council for the 21St century, appointed by the Ministry of Foreign Affairs (MOFA) and composed of representatives from academia, mass media, the private sector, diplomatic community, and NGOs published a report emphasizing the turning away from simple concepts measured by per-capita gross national product to environmental protection, local participation, consideration of indigenous social and cultural values, gender equality, NGO participation etc. This report also requested clearer directions as to what actions should be taken in particular contexts within Japan's ODA. It also advocated a more pro-active approach through jointly initiating activities with the recipient countries and hence establishing real partnerships.

This all guided major changes in Japan's ODA programme, in which the ODA administrative system was to become consolidated under the MOFA. Moreover, it was intended that the Japan International Cooperation Agency (JICA) would assume a central and co-coordinating role in the implementation of technical cooperation. In addition the Japanese government decided to prepare a five-year policy guideline and a country-by-country strategy, initially for the eleven major recipient countries including Indonesia (OECD 1999: p. 24ff). These measures were also regarded as an another significant development in Japanese ODA reforms in the second half of the 1990s in order to push for improved transparency and efficiency by concerned ministries and agencies.

11.2.1 ODA Charta

The Japanese government responded to qualitative and quantitative changes in the structure of Japanese development assistance at the beginning of the 1990s with the formulation of a philosophy for cooperation - the ODA Charta. The ODA Charta was aimed to serve as a guideline to strengthen the understanding for Japan's ODA programme domestically as well as in foreign countries allowing for more effective and more efficient implementation. During the meeting of the Japanese Cabinet on 30 June 1992, only a few days after UNCED, the ODA Charta was accepted. It was led by four fundamental principles:

1. Emphasis of the compatibility of economic and ecological development,
2. Avoidance of developing help for military purposes and/or to escalation of international conflicts,
3. Supervision of the military expenditure of the developing countries, the development and manufacture of weapons of mass destruction and the international arms trade,
4. Attentive observation of the democratization progress of developing countries, the introduction of the market economy and the guarantees of human rights - particularly freedom (MOFA 1996: p. 46ff).

The Charta was understood as a declaration of the goals and policies of Japan's prospective development cooperation. It contained decisive political elements like democratization, fundamental human rights and militarism. Principles became clear through the "compatibility of economy and ecology" as well as the "introduction of the market economy". The ODA-Charta made clear the urgent effort to orient Japan's development aid more politically in the future (Hashimoto 1995: p. 84ff).

For Japan ODA is an integral part of its nation's foreign policy and considered as an important diplomatic tool at the face of principally renouncing the goal of becoming a military power. Hence, Japan's ODA programme is essentially based on economic and diplomatic foundations, supporting the societal development of the recipient countries in order to contribute to their development needs and the national interests of Japan. The ODA Charta affirmed that many people still suffer from famine and poverty – aspects that cannot be ignored in any attempt to ensure stability and further development in the Third World.

Some development projects in the 1990s, however, were incompatible with the principles of the ODA Charta, which states the value of sovereign equality and non-intervention in domestic matters of recipient countries, but should not leave human rights aspects aside. One example was the resumption of the development aid for Myanmar, which is subordinated by a military regime until today. Also in the 1990s development aid for the People`s Republic of China, which substantially increased its military expenditure and provoked Taiwan militarily, and which suffocated the work of the opposition as well as the ODA for Indonesia are additional examples. In the 1990s the Indonesian government brutally suppressed all efforts that would enable East Timor to become independent and thus acted against Japan's ODA Charta. In this respect Japan's practical ODA infringed against the self-made Charta in many ways.

The comparison of suspending development aid to Nigeria in March 1994, which was governed by a military regime under General Sani *Abacha* from November 1993 until 1998 and did not show any attempts towards democratization during this period, with the tremendous growth of the development aid budget for the People's Republic of China[124] in the 1990s infers that Japan`s development cooperation was conducted by superior interests. From the view of political responsibles of Japan, this, deviating from the principles of the ODA Charta, seemed to be justified.

The investigation of the second and third phase of the policy-cycle will present whether this assertion can also be proven after a detailed investigation of projects transferring environmental technology as a part of Japan's transnational environmental policies to NICs, using Indonesia as a case.

The international and national discussion on the different levels of the environment problem, first institutional changes in organizing development aid as part of the transnational environmental policies as well as the listing of environment guidelines through the OECF and the adoption of the ODA Charta and following reforms have worked out a frame for Japan's eco-political ODA and thus demonstrated potential for alternative scenarios of transnational environmental policies. The deviation from the principles of the ODA Charta in the cases of China and Myanmar macro-analytically put the rating of development aid motives in the last rank of a hierarchy of the motives for strengthened eco-political involvement. Whether this assessment can also be confirmed after a micro-analytic investigation, with inclusion of the individual conditions of eco-orientated development projects through environmental technology transfer to

---

124 From US$ 1.05 billion in 1992 to US$ 1.479 billion in 1994.

Indonesia, will be shown by application of the suitable framework of actions through certain policies. As synthesis of micro-analysis and macro-analysis, which supplies knowledge about compatibility with certain guidelines like the ODA Charta and the recommendations of the ODA Reform Council, statements about the quality of Japan's transnational environmental policies through environmental technology transfer to NICs such as Indonesia can be made.

In response to the Agenda 21, Germany also supplemented its basic principles[125] of 1986 in 1996 with the Development Policy Concept. It identified poverty alleviation, environmental and resource protection and education and training as the three priority areas of German development cooperation. Similar to Japan Germany was also increasingly in the position to justify the use of taxpayers' money for ODA. Therefore, economic returns of the development cooperation programme entered into the political focus in the late 1990s. Nevertheless, moral considerations and a broad approach towards international security remained main motives behind Germany's ODA. The following five criteria regarded as conducive to successful development cooperation by Germany:

1. Respect for human rights,
2. Peoples' participation in political decisions,
3. Rule of law,
4. Introduction of a social market economy,
5. Development commitment of the partner government.

These criteria are not always applied with the same rigour in policy dialogues with partner countries; China has been frequently cited in this respect, although the German Government and Parliament have been more critical of the human rights situation in China than many other donors.

---

125 The Basic Principles of Development Policy of the Federal Government adopted in 1986, stipulate: "The aim of the German development policy is to improve the economic and social situation of the people in developing countries and to provide scope for their creative growth. It therefore helps meet the basic needs of the people and enable them to help themselves. It contributes to the development of a viable economy and social diversity as the prerequisite for the country's self-sufficient development. And it promotes regional co-operation and facilitates the integration of developing countries into the world economy." (OECD 1999b: p. 13).

12 Selection

Within the third step of the Policy-Cycle Model, the options for action identified during the policy initiation are assessed based on a detailed cost-benefit and impact analysis (Prittwitz 1990: p. 96). This implies the need for an activity programme and therefore of a defined goal, the necessary strategies and instruments to reach the respective goal. In consequence the selection also contributes to the clarification of values prioritized by the policy makers and provides additional indications of the actor's respective powers.

12.1 Basic Environment Law and Basic Environment Plan

In response to the UNCED in Rio, the Japanese Parliament adopted the United Nations Framework Convention on Climate Change (UNFCCC) and the Convention on Biological Diversity. Additionally, the Basic Environment Law (BEL) was enacted in November 1993 (Kühr 1996: p. 5). The BEL firmly defines the frame of fundamental principles for environmental protection, including transnational and global aspects.

Since the beginning of the 1970s, a new environment law providing guidelines had already been a top priority of the Environment Agency (Meves 1993: p. 176). With the growing interest of some LDP leaders in the environmental challenges, the EA addressed the topic of a new environment law in October 1992. To ensure the support of the majority of his party, Takeshita ensured that the chairman of the party's team on building politics, foreign policy, agriculture and trade were represented in the *Investigatory Committee on Basic Environment Problems* with Ryûtarô *Hashimoto* as chairman. However, the Ministry of Construction, the Ministry of Transport, MITI and the Keidanren[126] found that the introduction of an Environmental Impact Assessment would be a contradiction to existing cabinet decisions and voluntary initiatives of Japanese industry. They argued that such a law could lead to numerous legal cases e.g. against building projects. Furthermore, they argued that this law could hinder economic development and large building projects could be suspended indefinitely.

Because of the "Sagawa Kyûbin Scandal"[127] Takeshita's influence to receive support through his party for the adoption of the EIA and for environmental tax-

---

126 Japan Federation of Economic Organizations, established in 1946.
127 Takeshita could only take the position of Prime Minister with the help of Shin Kanemaru after Takeshita's initiatives towards the splitting of the Tanaka Faction. Through the Sagawa Kyûbin Scandal the connection of Kanemaru with organized crime became obvious.

es waned. Because of his decreasing support, the draft EA was turned down and a new bill was ultimately worked out with the sole objective to improve existing measures for environmental protection (Schreurs 1994: p. 20f.).

A similar draft was adopted by the government of Prime Minister Morihiro *Hosokawa* (05 August 1993 - 24 April 1994) on 12 November 1993. Predominantly, economic interests had decisively watered down the real goal of an environment law providing guidelines. The BEL describes fundamental measures, which were planned inclusively and was expected to contribute to the education of Japanese society but also to enable a development towards sustainability. With the adoption of the Basic Environment Plan (BEP) on 16 December 1994 on the basis of the BEL, Japan laid a base for its future environmental policy and emphasised three measures:

1. EIA can help to consider the environment in formulating and implementing policies of the government and/or in the formation of public works.
2. Investigations over economic measures like e.g. environmental taxes, which aim to consider environmental costs through economic activities, shall explain the consequences of environmental protection, economic development and other areas.
3. An active contribution for mankind should be the research of the natural and social sciences, the monitoring and the promotion of EST (Government of Japan 1995: p. 3).

It became obvious that the BEP of the Japanese government felt back on the standards of the Environment Agency in its draft for this Plan to consider in both domestic and foreign policies their efforts to actively support environmental protection. Whether the ecology had to step back, however, behind economic interests will be shown in the following by the transformation of principals into concrete policies.

12.2 Green Aid Plan

Japan enacted the Green Aid Plan (GAP) for the promotion of technical cooperation with developing countries. The GAP was proposed by MITI and since 1992 executed by 14 subordinate organizations including Japan External Trade Organization (JETRO), the Association of Overseas Technical Scholarship (AOTS), the Japan Overseas Development Corporation (JODC), the International Centre for Environmental Technology Transfer (ICETT), the Engineering Consulting Firms Association (ECFA),the New Energy Development Organization (NEDO) and private industry enterprises, which themselves were under NEDO's control (MITI 1995: p. 88).

The Green Aid Plan aimed to improve the environment of developing countries; it addressed air and water pollution, waste disposal and recycling as well as waste of energy. One characteristic of the GAP was that it linked political dialogue between MITI and the government of the recipient state. The project shaping was coordinated through JETRO[128]. However, the potential recipient country's application was consciously renounced at the respective Japanese Embassy; it was NEDO deciding which environmental technology is brought into action in the corresponding recipient country. As a result GAP met with the increasing interest of Japanese producers of environmental technology because it minimized costs and risks of the adjustment and testing of various potential export markets.

GAP projects were instrumental in solidifying long-term partnerships with the respective industry of the recipient state (Evans 1994: p. 40). Thailand, Indonesia, Malaysia, the Philippines and the People's Republic of China enjoyed priority (Maruko 1995: p. 8), although since 1994 an expansion of the GAP to other countries has been under discussion. With the exception of Malaysia, these above-mentioned countries could be found in the list of "Top-10-recipients of Japanese ODA" in 1994 and received altogether US$ 3.345 billion in bilateral ODA, which totalled one-third of the budget of Japan's bilateral development aid; the share of Asia totalled 57.3% in 1994. This concentration in East and Southeast Asia was frequently criticized by Japanese development aid politics, even though regional concentration was typical for nearly all donors in the 1990s. For instance the United States spent almost 50% of their development aid on Egypt, Israel and Russia in 1994; increasingly, Japan was also a top donor of development assistance for nations in Africa, South America and the Far East (Green 1994: p. 112). Still, Asia enjoyed supreme priority in the Japanese development cooperation. The ODA Charta says:

> "Historically, geographically, politically and economically, Asia is a region close to Japan. East Asian countries, especially member countries of the Association of South East Asian Nations (ASEAN) constitute one of the most economically dynamic regions in the world, and it is important for the world economy as a whole to sustain and promote the economic development of these countries. There are however, some Asian countries where large segments of the population still suffer from poverty. Asia therefore will continue to be a priority region for Japan's ODA."(ODA Charter, Ch. 3)

---

[128] The Japan External Trade Organization (JETRO) is a Japanese government-related organization that promotes mutually beneficial trade and investment relations.

The concentration of Japanese ODA in the rapid growing states of Asia in the 1990s strengthens the hypothesis that this was primarily due to the GAP's economic motivation as well as Japan's entire development cooperation. Besides the growing market for environmental technology, broader economic interests resulting from fast growing markets in Indonesia was also observed. Preventative environmental technology yields business savings for cost-intensive clean-up technologies possible in a way that these cost savings increases the investment volume of industry, not only for ET (US International Trade Commission 1993: p. 142f).

Likewise, the GAP can be perceived as Tokyo's intention to take a leading role in the transnational environmental policy in the 1990s as well as to compensate for the looting of natural resources from some recipients. There is an indication that these projects were identifying transnational environment problems as the climate change discussion assumed a more visible role in the international arena.

However, MITI identified the power supply issue as being essential in improving the environmental conditions (Yamamoto 1994: p. 14), and certainly remains essential even today. By 2010 a 150% increase in energy consumption compared to 1993 consumption was forecasted for Asia; between 1998 and 2008 energy consumption grew by an astounding 70% (The Economist 2010).

Thus, the GAP also played a particular role in securing energy sources for Japan; growing demand for less polluting energies like low sulphur coal and natural gas would have had dire consequences on Japan's production due to climbing prices. With the necessary desulphurization technology, China was expected to return to using coal high in sulphur due to its large stocks Likewise adjustments can increase the effectiveness of existing nuclear power plants and decrease the fear of possible accidents because of the relatively high safety standards of Japanese atomic energy technology as well as stabilize potential further interest in such energy sources (Murdo 1994: p. 16).

12.2.1 Green Aid Plan for Indonesia

As part of the Green Aid Plan, Japanese experts were sent to Indonesia in 1995 in order to develop cooperation for human resources. As experts of the energy and environmental sector, their duty was seen in the technical instruction of responsible Indonesian parties in the private sector; The Japan Overseas Development Corporation (JODC) was leading the realization of such cooperation with Indonesia. Likewise, in the implementation of JETRO, a mutual exchange of four experts had to take place to strengthen Indonesian knowledge for the plan on environmental protection measures as well as to ascertain ET know-how from Japanese experts enabling the Indonesians to improve ET use and effectiveness in Indonesian enterprises. Altogether in the mid-1990s 20 Indonesian engineers in industrial technology were trained under the supervision of the Japanese Association of Technical Scholarships (AOTS) in water protection, management of industrial waste and air protection in different enterprises. In addition the dissemination of knowledge from Japanese experts was organized through JETRO and AOTS introducing Japanese experiences with environmental technology in both the environment and energy sectors.

Cooperation between Japan and Indonesia had to be implemented through five different projects:

1. In the JETRO project *Basic Survey on Protective Measures for Environment* Japanese experts examined Indonesian energy and environment conditions to demonstrate existing problems and starting points for technical cooperation as a solution.
2. The ICETT *Eco-Phoenix Plan* project analyzed existing pollution, the responsible pollutants and formulated advice and recommendations.
3. A master plan for the use of coal that considered technical requirements and economic development if Indonesia undertook activities in the project entitled *International Co-operation for the Diffusion of Clean Coal Technology Research Program for an Environmentally Friendly Coal Utilisation System*
4. There was also the *Basic Research Project for Data Base Development and Energy Rationalisation Planning in Industrial Sector* as well as the
5. The NEDO *Basic Research Project for Energy Rationalisation Planning in Industrial Sector Co-operation for Efficiency Diagnosis* project which identified efficient measures for energy consumption in Indonesia.

These projects were carried out to strengthen the research cooperation between Japan and Indonesia in addition to various NEDO pilot-projects with various Indonesian institutions. Consequently, a common pilot-project – *Development of Simple Purification System for Industrial Waste Water* produced national economic cost-effective methods for the clarification of sewage production. The project entitled *Research Co-operation Concerning Conservation and Sustainable Use of Biological Diversity* explored technology that effectively protects natural resources e.g. in the tropical rain forest. Common research under the *Development of Laser Radar for Environmental Measurement* facilitated the observation of urban air pollution; four further pilot demonstration projects were also included the Green Aid Plan for Indonesia in 1995.

Two projects addressed increasing the effectiveness of energy consumption in steel production and pre-shredder plants in cement mills. These projects were entitled Verification Research on a Material Preheater for Electric Furnaces in the Steelmaking Process and Improvement of Efficiency in Large Size boilers for Industrial Use - Demonstration Project of Pre-grinder Equipment. A further demonstration project – International Co-operation for the Diffusion of Clean Coal Technologies - Development Program for an Environmentally Friendly Coal Utilisation System devoted research to the manufacturing of briquettes. The chief objective of the fourth project was to install room boilers for hot water processing (MITI 1995a).

The implementation of numerous developing projects in Indonesia is done through NEDO, a semi-governmental organization (Incorporated Administrative Agency) still active today since its establishment in 1980. NEDO's budget stems primarily from the METI[129] and therefore requires endorsement of the METI for its projects and financial plans[130]. Likewise, the fact speaks for the thesis numerous staff from NEDO, JETRO and ICETT were lent temporarily by private industry for very special and concrete projects in the 1990s. Thus, in the 1990s Japan's ODA was narrowly linked to the commercial interests of the private sector protagonists and the strategic economic agenda of different ministries like MITI, the Ministry of Finance and the Economic Planning Agency (EPA).

At the same time it was increasingly requested of Germany that it use a portion of its ODA to support its economic efforts in NICs, including Indonesia. The German economy was criticized for not making use of the opportunities in the dynamic states of Asia and therefore lost potential market shares (BMZ 1993: p. 2). In this sense contrary to Japan, Germany's ODA was less commercial-oriented and focused more on the sustainability paradigm instead of short-term economic successes (BMZ 1997: p. 46). In consequence the GTZ started to develop new means for cooperation between the private sector and institutions in the sector of environmental technology transfer following an expert meeting in 1997 on technology transfer and increased technological capability (GTZ 1997). Moreover, upon initiation of the German Minister for the Environment, Klaus Töpfer, the establishment of the Centre for the International Transfer of Environmental Techniques (ITUT) in 1996 served as a joint initiative between

---

129 As part of the central government reform in 2001 the former Ministry of International Trade and Industry (tsūshō-sangyō-shō) (MITI) was reorganized to the Ministry of Economy, Trade and Industry (keizai sangyō shō) (METI).
130 Today, the Ministry of Environment co-administers NEDO's Kyoto Mechanisms Credit Acquisition Programme.

politics and business but also as a response to the Germany's decreasing stake in the global environmental technologies market.

In direct comparison with Japan, Germany started efforts to increasingly commercialize its transfer of environmental technologies to NICs in the latter half of the 1990s. The influence of private enterprises on Japan's ODA already began in the early years of rapid economic growth in the mid-1950s and continued almost unchanged in the organization structures of Japanese ODA in 1990s.[131] Also, after the amendment of the ODA Charter in 2003, which attempted a strategic new orientation of Japan's development cooperation, Japanese economic growth and securing resources for Japan's production chain remained key-targets.

12.3 National Action Plan for "Agenda 21"

The adoption of Agenda 21 through the Rio-Conference also required signing parties to prepare national action plans. During the G7 summit in summer of 1993 in Tokyo, the seven leading economic nations agreed to present their respective programmes by the end of 1993 and to forward them to the UN Commission on Sustainable Development (CSD).[132]

Japan's action plan presented a first step to protect the transnational and global environment. It emphasized its firm will to follow the demands of Agenda 21 and underlined its effort to take a leading role in transnational environmental protection.

On 159 pages, subdivided into 40 chapters assigned to four different sections - *social and economic dimensions, protection and management of resources for development, strengthening of the role of important groups*, and *ways of implementation* - concrete plans for the realization of Agenda 21 are presented. The 34th chapter is entitled "Transfer of Environmentally Sound Technology (EST), Cooperation and Capacity-Building".

---

131 One cabinet decision (kakugi kettei) under Prime Minister Shigeru Yoshia (19 October 1948 -09 December 1954) with the title kakugi kettei – Ajia shokoku ni tai suru keizai kyōryoku ni kan suru ken (Policy on Economic Cooperation with the Countries of Asia) says that economic cooperation is mainly taking place through initiatives of private industry and only requiring support from the government. Various bodies like Ajia keizai kondankai (Asian Economic Deliberation Council) with well-known representatives from private industry integrate their views and goals in the international developing cooperation (Arase 1994: p. 173 ff.).

132 The CSD is in charge of the follow-up on Agenda 21 according to UNCED. It belongs to the UN Economic and Social Council (ECOSOC).

Considering the lack of appropriate environmental technologies and because of the lack of know-how for effective utilization of existing technology in the 1990s, it was an urgent goal to improve effective use of environmental technology through training following this action plan. These and further measures had served in particular in NICs resulting out of the speedy industrial development. Simultaneously, original factors and specific conditions in the different regions had to be documented yielding more effective formation of technology transfer, which were regarded important for the use of ET. Japan supported the development of such a database through the Osaka and Shiga based International Environmental Technology Centre (ITEC) under UNEP's Division for Technology, Economic and Industry. Furthermore, Japan wanted to make use of the experiences and data of public and private institutions; in this sense Japan contributed actively to the formation of an international network supporting the transfer of environmental technology through UNDP (Government of Japan 1994: Chapter 35, Paragraph A).

To guarantee sustainable development with reduced ecological damage, Japan originally planned to expand its budget for the Green Aid Plan as well as the number of recipient states in the mid-1990s. Despite a decrease in Japan's ODA in the second half of the 1990s there was growth until the end 1990s.

Simultaneously, it was planned to increase the number of dispatched experts for development, the number of trainees from developing countries and increase the cooperation agreements for EST research. Criticisms of Japan's development aid politics frequently referred to the number of civil servants in the area of developing cooperation (Yamada 1995b: p. 13).

JICA personnel only increased from 994 to 1,158 employees between 1974 and 1994, although the budget increased in this period from JP¥ 27.2 billion to JP¥ 172.6 billion. At the end of the 1990s, the German GTZ employed approximately 1,300 which corresponded only to the number of staff at GTZ headquarters in Eschborn. The number of experts from other semi-governmental institutions like the Engineering Consulting Firms Association (ECFA), ICETT, the Japanese Development Institutes (JDI) and private consultants like ExCorporation, which were increasingly involved in JICA projects, climbed however from just 500 persons in 1974 to 3,098 in 1996 (JICA 1996: p. 10). The German GTZ only dispatched approximately 1,500 experts at the same time, roughly 50% less than Japan, although more than 7,000 persons were working for the GTZ worldwide at that time.

Taking the budget for environmental projects of JP¥ 21.873 billion in 1994 into account, which equalled one-eighth of the JICA's total budget, the dispatch of 220 environment experts seems very low and a perceptible increase of this quota is desirable (JICA 1994: p. 13f).

Furthermore, Tokyo aimed at supporting various projects for the improvement of the environment in developing countries as well as for the decrease of transnational and hence partly global environment problems – including geological investigations for the prevention against natural disasters like earthquakes and volcano outbreaks, financial and content support of measures for sustainable development against traffic problems, and contributions to the multilateral fund for the protection of the ozone layer as well as the construction of the EST database (Government of Japan 1994: Chapter 35, Paragraph B).

Environmental centres in NICs like Indonesia and China were expected to serve as training institutions for human resources that are (i) responsible for environmental problems and (ii) in the position to find and influence sustainable solutions. Additionally, they had to set up cooperation plans for preventative measures against air pollution and more ecologically friendly transport systems (Ibid.: Paragraph C).

The Japanese National Action Plan on Agenda 21 prescribed contributing to the exchange of information and data among the environment research and environment training centres in China, Thailand and Indonesia, financed through Japanese ODA, as well as UNEP/IETC. This multilateral research was also supported through the Consultative Group on International Agricultural Research (CGIAR) and UNESCO (Ibid.: Paragraph D).

In light of the fact that numerous technologies had to be modified for the individual needs of the corresponding countries, Tokyo planned supporting Japanese enterprises, considering the traditions and customs of the developing country. Thereby, agreements had to apply above all to the license award with enterprises which transfer easily integrateable technology. Moreover, simpler trade conditions had to support initiatives for the transfer of environmental technology through private enterprises (Ibid.: Paragraph G).

With the Basic Environment Law and the Basic Environment Plan, Japan's Green Aid Plan and the National Action Plan for Agenda 21, Japan developed programmes and instruments in the 1990s which were expected to lead to sustainable development. The above analysis presented various motives like the securing of raw material resources for Japan's production chains, the solving of transnational and global environment problems, the opening of growth markets for Japanese environmental technologies in NICs but also sustainability-oriented development in developing countries. It became clear that Japan was willing to assume increasing responsibility through reinforcing cooperation on bilateral and multilateral levels. Whether and how Japan was successful will be demonstrated in the following section.

13 Implementation

In the fourth stage of the policy-cycle, the implementation, political intentions are converted into measurable deeds. The previously agreed on action programme lays down categorically the course of activities for the political administrative system. And the laws and plans of action form a mean of orientation for carrying out a certain policy. The specific "fulfilment" of the prescribed framework programmes takes place in the phase of implementation, so that often even decisive changes in content can occur. This phase implicates therefore a second process of programme development conceived in practice, through which the hierarchy of the motives and values becomes more apparent (Schubert 1991: p. 74f), also contributing to a better understanding of the powers driving action, and the inter-linkages between the different actors involved.

13.1 Institutional Framework

The various aspects of environmental problems make transnational environmental policy through the transfer of environmental technology to NICs a complex task. In the 1990s environmental problems were regarded as restricted to local regions. Because of Japan's environmental history local administrations in Japan (*chihō seifu*) had developed a special skill in dealing with environmental problems; private industry contributed to this through the development of the necessary technologies. For this reason cooperation of local administrations with the central government (*chūō seifu*) was and is still just as essential a cooperation with private industry, in order to transfer technologies to where they were needed.

13.2 Consulting Companies

Because of the increasing demand for environmental policy aid as a part of Japanese ODA-policy, JICA and EA started to increasingly employ consultants to collect information and formulate projects in the 1990s. Even today, Japan is one of the only countries substantially represented by semi-governmental organizations such as NEDO or private consultants such as EX Corporation in the work of multilateral environmental agreements such as the Basel Convention on the Control of Transboundary Movements of Hazardous Wastes and their Disposal.

Consulting agencies (*konsarutanto gaisha*) act as private companies, maintain close contacts with private industry and are therefore not necessarily restricted by the diplomatic entanglement of the state aid organizations (Forrest 1991: p. 27). They are able to give the governments of the donating/granting country and recipient concrete suggestions. Moreover, through the assistance of the consulting agencies, the governments of the recipient countries are able to apply formally but also successfully for development cooperation at the Japanese embassy. Thus, the sovereignty of the receiving country remains intact, while the interests of the Japanese are already included in the formulation of the project. Here the consultants profit from their close contacts with the government officials of both sides, so that the consulting companies or their subsidiaries are entrusted with further work on the project (Schmitt 1991). On the other hand this system leaves plenty of loopholes for corruption and the use of official ODA funds for private interests.[133] In the mid-1990s personnel of the consulting companies started to recruit more and more young, highly qualified experts worldwide, whose experience and innovative thinking was expected to give Japan's environmental aid a completely new dynamism.

In the 1990s the Consulting Company EX Corporation was commissioned by both the EA and the JICA to analyze the environmental conditions and problems in Indonesia, to analyze ODA environmental projects of different national and international aid organizations and to develop models and concepts for Japan's future environmental cooperation with Indonesia.

In March 1992 the 482 page study *Indonesia - Resources and the Environment. Toward an International Co-operation for the Environmental Management* was completed. In this study EX Corporation provides a clear and complete picture of the environmental problems in Indonesia, which have been caused not only by rapid economic development but also by the export of natural resources to Japan (EX Corporation 1992: p. 253). Japan imported 94% of its plywood and 14% of its crude oil from Indonesia.[134] Besides its mineral and energy resources, Indonesia also exported large quantities of lobster and tuna fish to Japan; it also provided Japanese industry with a lot of raw materials for the computer and car industries.

Therefore Japan had and still has today a special interest in and responsibility for Indonesia's further development. EX Corporation drew attention in its study to three areas for possible cooperation between Japan and Indonesia:

---

133 Reminiscent of the arrest of JICA's Director of the Agricultural Division in 1986 who received financial presents from a consulting agency for projects in Africa.
134 The demand for plywood as a building material is higher than in other industrialized countries caused by the large amount of wood in building and the demolition of houses after 20-30 years.

1. Protecting the natural environment
   A complete national development plan for Indonesia had to take account of the various aspects of urbanization, industrialization, transport, housing and urban infrastructure etc. This means that environmental data and information had to be collected and processed. EX Corporation gave priority to a plan for environmental management and effective use of natural resources. Also, reforestation, in which Japanese technology and experience could be used, gained special significance. In the eyes of the consultants an additional important measure was the training of Indonesian experts in the areas of environmental research and management.

2. Control of Pollution of the environment
   Thanks to its experience gained during the years of fast economic growth, Japan developed efficient cleaning technology and clean technology in the petrochemical industry, power stations and other factories, which it transferred to Indonesia in order to protect the environment. Similarly, Japanese experience could be of assistance in formulating a policy on the reduction of pollution by small and medium-sized companies. EX Corporation emphasized the importance of monitoring technology in order to keep an eye on pollution.

3. Request for development aid to the Japanese government
   Japan acceded to the request from the Indonesian government and promised support in the founding of an Environmental Management Centre. In addition Japan committed supporting further environment-orientated programmes, particularly in water management with technology and know-how (EX Corporation 1992: p. 482ff).

At first sight the role of consultancies in Germany's transnational environmental policy is less prominent as in Japan. In comparison with Japan Germany does not send consultancies to observe and become actively involved in the work under multilateral environmental agreements. However, the large amount of consultancy studies assigned and endorsed by governmental organizations illustrating an enormous economic potential for Germany through transferring environmental technologies, though not necessarily referencing the sources of this information or disclosing how these data was generated, illustrates two aspects: First, similar to Japan a rather nearness of the German government to consultants in the sphere of environmental technology transfer, in order to benefit from their technological skills and connections to the private sector, usually owning the technologies. Second, similar to Japan the support of the German environmental technology business has the special attention of the government. Both,

Japan and Germany, see the support of their respective national business as a key-justification, also against the continuously arising questions associated with their development cooperation. Hence, the private sector, whether consultants or owners of the technology to be transferred, are commonly regarded as key-actors in this field of transnational environmental policy, whereas this does not give any evidence about an adjusted role of governments and the nation-states.

Some of the suggestions from EX Corporation were implemented by local administrations and were incorporated in follow-up programmes of the central government in Tokyo, which will be analyzed in the following.

13.3 Environmental Technology Transfer by Local Organizations

In the 1990s the transfer of environmental technology to NICs was mainly conducted by two local governments, which made particularly bitter acquaintance with pollution of the environment as a result of fast economic growth. Not only Yokkaichi, where ICETT was established, but also Kitakyushu, where the Kitakyushu International Training Association (KITA) has its home, had many victims of extreme air pollution in the 1960s. The city and prefecture of Osaka invited the Global Environment Centre Foundation (GEC) to take up residence in Osaka. The International Lake Environment Committee Foundation (ILEC) accepted the invitation of the Shiga prefecture and settled in Kusatsu. Both GEC and ILEC formed the pillars of the International Environmental Technology Centre (ITEC) of UNEP in Osaka and Shiga.

13.3.1 Kitakyushu International Techno-Cooperative Association

Early in the 1980s the Kitakyushu International Training Association was founded by the city of Kitakyushu, the prefecture of Fukuoka and private firms in the region. It was renamed Kitakyushu International Techno-cooperative Association (KITA) in August 1992. At the beginning KITA's main objective was to support JICA training programmes in the iron and steel industry; since then KITA has been cooperating closely with JICA's Kyushu International Centre in creating training programmes for the transfer of technology.

For many years KITA concentrated its activities on the transfer of industrial technology; since the establishment of the KITA Environmental Cooperation Centre, a shift towards environmental problems took place. In the 1990s KITA was supported in its work by more than 200 organizations, including 139 private firms, 16 working parties at universities and 27 government organizations at the local and regional level (Yamamoto 1994: p. 18f). In order to carry out its work,

KITA cooperated with institutions such as EA, MITI, JICA and UNEP; with an annual budget of only JP¥ 1.5 billion in the mid-1990s, KITA mainly used volunteers for the coordination of its activities, the majority being retired from private industry. Until mid-1996 approximately 230 people were trained in environmental programmes and environmental experts were sent to China, Korea, Indonesia and Singapore. In 2008 KITA trained approximately 500 persons, the large majority of them from Asia (KITA 2009); similarly, KITA made use of research and development in order to draw conclusions for developing countries thanks to Kitakyushu's experience with air pollution in the years of swift economic growth.

13.3.2 International Environmental Technology Centre

In 1992 UNEP decided to establish an International Environmental Technology Centre (IETC) based in Osaka and Kusatsu (Shiga Prefecture). The expertise generated by the Global Environment Centre Foundation (GEC) through dealing with environmental problems in Osaka was transferred to the newly established IETC. GEC was concerned with urban environmental problems such as air pollution, waste management, sewage treatment, protection from noise etc. In addition the International Lake Environment Committee Foundation (ILEC) supported IETC in its work by contributing its expert knowledge on fresh water management.

IETC planned to act as a catalyst in order to make the transfer of environmental technology easier - this included the transfer of ET not only from developing nations but also from industrialized nations like Japan. In the first years of its operation, IETC understood its role as a mediator for the cooperation between manufacturers and users of ET (Uitto 1994: p. 173) by connecting all interested parties in one network.[135] At the same time IETC aimed at improving the capability of human resources involved in the utilization of environmental technologies in developing countries and NICs to support nations which strived for a more sustainable development through the utilization of environmental technology according to Agenda 21. However, it was only in April 1994 that the IETC began its work after the opening of its institutes in Osaka and Kusatsu.
In the 1990s IETC activities included:

- Improving the knowledge base by undertaking overviews of ETs in priority sectors and implementing partnership arrangements with key ET information providers,

---

135 See IETC homepage http://www.isda.or.jp/kansai/k1_e.html – 10 September 2010.

- Making information accessible by developing and maintaining an ET information system (maESTro) that meets priority needs reflected in environmental conventions and identified by governments, particularly in Africa and small islands developing states,
- Strengthening and improving capacity in developing countries and countries with economies in transition to adopt and use ETs in the areas of freshwater and urban environmental management. (UNEP 1998: p. 6).

Nevertheless, in the 1990s the main focus was on the collection and dissemination of ET, whereas the urban and freshwater environmental management issues were less visible. In addition IETC was criticized for lacking a clear vision and focus for its activities, resulting mainly in stand-alone projects (Schelleman 1998). 90% of these activities were financed by the Japanese Ministry of Foreign Affairs, which made it highly dependent on one single source with all associated risks due to budgetary problems in the overall Japanese budget.

In the latter half of the 2000s, IETC continued implementing its activities and programmes in three areas:

1. Water and sanitation,
2. Sustainable consumption and production with a special focus on waste management and
3. Disaster prevention and management.

Though the application of ET remained the underlying thrust in all three areas, IETC changed its focus to more ground-level application of ETs and capacity building in developing countries, building on the expertise cultivated in the 1990s.

Today, IETC focuses on waste management as a response to Japan's 3R activities; this has led to less emphasis on water and sanitation. IETC, similar to all United Nations (UN) Organizations in Japan, is facing substantial funding cuts. In the past years the appreciation of the JP¥ against the US$ resulted in a substantial decrease of the donations from Japan to the UN.

13.3.3 International Centre for Environmental Technology Transfer

As a reaction to numerous requests from abroad for the transfer of know-how of Yokkaichi, the International Centre for Environmental Technology Transfer (ICETT) was founded in 1990 as a non-profit organization under the umbrella of MITI (Nose 1994: p. 21). As an element of Japan's international development aid, ICETT had and still has the task of making contributions to the "healthy industrialization" of developing nations (ICETT 1995: p. 2).

With budget of JP¥ 6 billion provided by the government of the Mie prefecture, the town council of Yokkaichi along with 200+ companies, established ICETT headquarters in the mountains near Yokkaichi. In the mid-1990s ICETT activities were concentrated in four main categories:

1. Development of Human Resources (Training)
   The development of human resources took place in the form of two types of training: a) in Japan, for which trainees were invited to come from developing nations and b) abroad, to which Japanese experts were sent out. Both kinds of training programmes aimed at improving the technical capabilities of trainees and simultaneously train them in ecological thinking.

2. National Studies
   A complete examination of the environmental conditions in the developing countries was considered necessary in order to collect enough information and data which can guarantee a correct approach to a problem. As much as possible had to be generated in order to promote mutual understanding for a close cooperation with ICETT and to identify the technical capabilities and other related factors of the developing countries. The results of these national studies were reflected in ICETT's training programmes.

3. Research and Development
   An important aim in the work of the ICETT was achieved when suitable environmental technology was transferred to companies in the developing countries and especially NICs and installed so that they worked on a stable basis. From this point of view, the development of suitable technology was a major activity of ICETT.

4. The Dissemination of Information, Data, etc.
   In order to increase the significance of environmental protection in the developing countries, ICETT regarded the active dissemination of information and data as well as of suitable measures against pollution significant for its activities. Seminars, training, symposia and exhibitions had to serve as fora in order to communicate the experiences of ICETT and the city of Yokkaichi.

The geographical situation of ICETT was and is until today is essential. Thanks to the support of the local petrochemical industry, demonstration projects, practical experiences and experimental laboratories were available, which demonstrated and improved different environmental technologies. At the same time close cooperation with the academic community in the Mie prefecture made it possible to provide scientific support to the activities of ICETT and to develop new environmental technologies. Specialists from MITI, the local government and various companies were granted sabbaticals in order to work at ICETT.

13.3.3.1 Palembang Eco-Phoenix Plan Project

Because of its industrial development and the rapid growth of the urban population, ICETT selected Palembang in Southeast Sumatra for its Eco-Phoenix Plan. In view of an estimated 1.5% annual population growth rate until 2000 in South Sumatra (ICETT 1995: p. 1) and in light of a higher proportion of mining and industrial sector activities and a lower proportion in the agricultural sector, this project was expected to contribute to the reduction of industrial pollution. Similar to Yokkaichi, Palembang had a high concentration of oil refineries, chemical fertiliser production plants and power stations, which were growing at a very fast rate and which were expected to lead to severe environmental problems (ICETT 1994: p. III).

The report of the study group came to the conclusion that the natural environment of Palembang was influenced primarily by natural phenomena; the proportion of pollution caused by humans was relatively small. For instance, 50% of the organic substances in the Musi River were of natural origin and had been brought in by surface water. The large amount of smoke was produced by the seasonal bush fires. Nevertheless, the study emphasised that these environmental conditions could change quickly as a result of rapid economic and industrial development and effective preventive measures had to be taken.

Based on the monitoring results of the PROKASIH[136] (Program Kali Bersih - Water Cleanup Program), this study contained environmental recommendations for the crumb rubber industry situated in Palembang in addition to the oil refineries. A complete environmental system for these branches of industry was based on three components, according to the Eco-Phoenix Plan:

1. Comprehensive monitoring system to detect pollution, determine the pollutants and take counter-measures,
2. Creation of different institutions to protect the environment,
3. Development of human resources by the relevant training. (ICETT 1995: 96ff)

In contrast to the representatives of the oil refineries, those responsible in the crumb rubber industry showed great interest in improving the quality of their

---

[136] BAPEDAL (Badan Pengelolaan Dampak Lingkungan – Environmental Impact Management Agency) analyzed by order of the Indonesian government three rivers on Java and one on each Kalimantan and Sumatra (the Musi River). This abbreviation from Bahasa – PROKASIH – is accepted in the international literature.

effluent problems by using economically more favourable and more suitable environmental technology. The oil industry showed itself to be less conscious of the effluent problems and therefore would not or could not fulfil an important precondition for support by ICETT. Considering the environment-orientated development aid by Japan, the conscious readiness of the receiving organization played a key-role. This readiness also helped in finding the most effective environmental technology, which could combine not only local resources but also incorporate external input (ICETT 1995: p. 101).

The study provided recommendations for the sewage water treatment of the crumb rubber and oil industry, support in the sewage water treatment of various small and medium-sized companies, guidance for the integration of environmental aspects in announcements of industrial areas as well as a four-step model for the development of an effective monitoring system, organizational changes in administration and the development of capability. For this reason the Eco-Phoenix Plan was expected to serve as a reference for improvement of the environmental situation and as reference for sustainable environmental protection in the Palembang region (Ibid.: p. 105).

A special support of the Japanese private sector through the Palembang Eco-Phoenix Plan materialized, whereas the interests of the Japanese environmental technology companies were already included in the formulation of the project. Here, companies profit from their close contacts with ICETT and therefore also with governmental officials on both sides, so that the companies and/or their subsidiaries could be assigned further work in the Palembang region. Whether or not this is plausible remains unanswered as it is either not documented or inaccessible to the author. Moreover, close ties with the private sector must not necessarily be contra-productive to sustainable development; in order to assess a possible negative impact, a thorough analysis of the project region and respective changes would be required. This would go beyond the scope of this study, though this study attempts to be as holistic and comprehensive as permitted.

13.3.4 Centre for International Transfer of Environmental Techniques

In 1999 a study by Schitag Ernst & Young Consultancy analyzed Germany's environmental technology transfer to non-industrialized countries (UBA 1999). The only institute mentioned that was comparable to Japan's local initiatives through KITA, IETC and ICETT was the Leipzig based Centre for the "International Transfer of Environmental Techniques" (ITUT). ITUT was established in 1996 with the objective to strengthen the technical, scientific and political cooperation in environmental protection between Germany and its partner countries around the world as a follow up to the Rio Conference and the UNFCCC COP1 in Berlin. ITUT was a joint initiative of the German Federal government, the German trade and industry and the government of the German Federal State of Saxony. The ITUT GmbH, a company with limited liability, was in charge of advising companies on possible markets and export possibilities for EST, whereas the ITUT Verein, a registered society, was in charge of further developing the cooperation between politics and the German trade and industry, particularly by establishing a well functioning network of information. The initial funding of EUR 5 million for ITUT was from the German Environment Foundation (DBU) and was to last seven years.

The German Ministry of Environment (BMU) was keen on an active involvement with the German Chamber of Industry and Commerce (DIHT) as the central organization for 80 Chambers of Industry and Commerce in Germany as partner of the ITUT GmbH and founding member of the ITUT Verein. However, the employees of the DIHT were already rather active in the field of environmental technology transfer.

DIHT was also in charge of the coordination of the Environmental Area Managers at the German *Außenhandelskammern* (AHK), the members of the worldwide German Chamber Network, which still functions as a platform for the promotion of German environmental technology to important partners abroad and also assisting German companies in entering new markets. Wilhelm *Kulke* reported that it was originally planned to bring these Environmental Area Mangers under ITUT; this never came to fruition. The German Association of Local Machinery and Industrial Equipment Manufacturers (VDMA) and the German Association of the Chemical Industry (VCI) continued to develop their environmental technology related activities separated from ITUT. Over the years and contrary to the Japanese institutes, ITUT also did not succeed in establishing the intended close cooperation with industry players. Moreover, and again contrary to the Japanese approach, ITUT was not aiming to train environmental technology experts. Finally, various management failures led to the termination of ITUT after the initial funding through the DBU was depleted. Still today, the work of ITUT has not been integrated into any other institute.

All interviewed experts on German's environmental technology transfer such as Horst *Breier* (BMU), Bernd *Kitterer* (ITUT), Wilhelm *Kulke* (ITUT & DBU), Alois *Schneider* (BMZ), Horst *Pohle* (UBA) and Helmut *Schulz* (BMBF) emphasized the large diversity of local and regional activities on environmental technology transfer. Nevertheless, a mapping of these activities and experiences does not exist, illustrating a weakness in the German system.

13.4 Environmental Technology Transfer by National Agencies

13.4.1 Japan International Cooperation Agency Projects

Until 2003 the Japan International Cooperation Agency (JICA) was a semi-governmental organization under the jurisdiction of the Ministry of Foreign Affairs. In the 1990s JICA handled more than half of Japan's technical cooperation activities and about 60% of the capital grant aid. Today, JICA is an independent governmental agency and has become the largest bilateral development aid organization in 2008 after a merger between JICA and that part of the Japan Bank for International Cooperation in 2008 (JBIC) which extended concessional loans to developing countries. JICA is also responsible for administering Japan's non-diplomatic part of grant aid which is currently under the jurisdiction of the Ministry of Foreign Affairs. In consequence all three major Japanese ODA components—technical cooperation, grant aid and concessional loans—are now managed within JICA.

13.4.1.1 Integrated Air Quality Management

Jakarta, the capital of Indonesia, has been victim of intensive air pollution as a result of increasing motorization, urbanization and industrialization. In addition to the automobile traffic, there are dust emissions by thermal power stations in the North and cement works near Pulogadung, Tangerang and Bekasi regarded as the main causes of the deteriorating air quality (World Bank 1994: p. 67ff). Since 1970 various organizations such as the Agency of Meteorology and Geophysics, the Indonesian Health Ministry and the Institute for Urban Planning and Environment have been monitoring the air in Jakarta. Air quality management studies have been carried out by different donors of aid - among them the Third Jabotek Urban Development Project of the World Bank.

Notwithstanding the large number of studies, there was not enough effective implementation of measures against air pollution. For this reason the Japanese government responded to a request by the Indonesian government to identify the air polluting facilities and industries by conducting a detailed examination, to decide on a control strategy and to train Indonesians in this matter as well as to transfer the necessary technology. In 1994 JICA was asked by the Japanese government to carry out these tasks. While taking the socio-economic conditions into consideration, an air pollution control strategy was expected to be developed only by the year 2010. In addition examinations and analyses on the present state of the socio-economy, nature, meteorology, air quality and air pollution were to develop a concrete plan of action by the year 2000 (JICA 1995c);

the final report was planned for October/November 1996. However, the members of JICA, even at the beginning of their activities, came across unexpected difficulties in their work with Badan Pengendalian Dampak Lingkungan (Environment Impact Management Agency = BAPEDAL), about which they did not want to go into further detail with the author; the project was satisfactory completed in 1997.

13.4.1.2 Environmental Management Centre

In August 1993 the Environmental Management Centre (EMC) was opened in Serpong, in the periphery of the Indonesian capital, Jakarta, in cooperation with the Japanese government. It was subordinate to BAPEDAL, the implementing arm of the Indonesian government for the area of environmental management.[137] BAPEDAL was put in place in June 1990 by a decree from President *Suharto* and its 120 permanent employees and 50 foreign advisers were responsible for setting up a national laboratory system for creating an information system as well as for developing training establishments focusing on the field of environmental management (EMC 1995: p. 2). The EMC was intended to support the work of BAPEDAL, particularly in the areas of laboratory and monitoring technology (JICA 1994d: p. 25). On the Japanese side, JICA functioned as executive agency of the government. In the years 1991 and 1992 Japan provided in total JP¥ 2.687 billion in the form of grant aid (MOFA 1996). From 01 January 1993 until 31 December 1997 JICA offered the EMC project-type technical cooperation, which included the sending out of 15 Japanese experts, providing equipment and materials with a total value of JP¥ 90.603 million for the EMC as well as training 8 persons in Japan (JICA 1995a: p 114ff). Indonesia was responsible for the operational costs of the EMC.

The main task of the EMC was to monitor the environment and to improve the local laboratories, which have been entrusted by 27 local governments. The EMC was expected to provide assistance by identifying pollutants and thus support BAPEDAL with suitable legal means in improving the quality of the environment in Indonesia. The latest technology could be used in laboratories in order to examine water and air quality and the training of environmental experts who could work for BAPEDAL in the future was another identified requirement (JICA 1993: p. 50). With the financial support of World Bank, the plan was to contract 48 environment experts by BAPEDAL in March 1995; unfortunately, no qualified persons were found to fill these vacancies. The deficit in experienced engineers and technicians dealing with environmental technology was

---

137 According to Masahiro Ohta during an interview on 23 June 1995. Mr. Ohta was JICA's Chief Adviser for the Environmental Management Centre.

emphasized in a joint study by the EA, the Overseas Environmental Cooperation Centre (OECC) and BAPEDAL as the decisive cause of the lack of effective measures for environmental protection in Indonesia (EA, OECC and BAPEDAL 1994: p. 5).

Also, the setting up of an extensive environment data bank belonged to the tasks of EMC. Besides the education of Indonesian government employees, the collection of extensive environment data was the second basic element in the EMC workplan. This would stimulate an environment and development policy aimed at effectiveness, to lay down environment standards and to comprehensively incorporate environmental aspects in future planning (BAPEDAL 1994). At the same time the EMC, with its 68 employees, was the central point for the environment control centres in Indonesia at large in the 1990s, which, similar to the "BAPEDAL Regional Monitoring Capacity Development Project", were implemented and financed by loans from the OECF and/or other foreign development aid organizations (JICA 1994c: p. 52ff).

The German GTZ was comparable with JICA, though a federally-owned limited liability company and hence not a governmental organization, mainly operating on behalf of the German Federal Ministry for Economic Cooperation and Development (BMZ) though also assigned by other governmental organizations, as well as the European Union, the World Bank, the UN etc. In 2007 about 81.3% of its turnover was only from the German government; in the 1990s the GTZ increasingly concentrated its work on securing an ecological basis in Germany's partner countries with sustainable development as an overarching principle. Hence, the environmental-sound planning and evaluation of its projects, the integration of environmental and resource protection in its focus areas, the development of country-specific environmental programmes, and the planning and implementation of concrete environmental protection projects were among the GTZ activities. In Indonesia it supported the development of environmental capability for decision makers. Moreover, GTZ provided an information and support service for appropriate technologies named ISAT, also to increase the technological competence in its partner countries. A pilot-project entitle *Strengthening Environmental Technological Capability (ETC)* aimed at developing the initial stages for environmental technological capability in Indonesia, India and Thailand through elaborating proposals for an improved and harmonized interaction among states, industry, research and education, by integrating the private sector as a partner in the effort (UBA 1999: p. 41ff). This pilot-project did not have the expected success at the end of the 1990s. This was also because the GTZ approach was perceived by small and medium-sized enterprises (SME) as an attempt to take over consultancy and coordination and an incompatibility of the approaches of the GTZ and the German SMEs. Today

GTZ, now part of the GIZ, is mainly viewed as a consultant and hence actively competing with other actors in this segment, though from an exclusive position with the substantial backing through the German government.

13.4.2 Overseas Economic Cooperation Fund Projects

In fiscal year 1994, the OECF granted credits with the total value of JP¥ 13.136 billion for four environmental projects. All projects had an interest rate of 2.6% and a period for repayment of 30 years. All Japanese credits for Indonesia amounting JP¥ 157,966 billion were without constraint, i.e. the orders for the project being financed did not have to be placed with Japanese companies (OECF 1995: p. 141). Altogether the proportion of loans without constraints amounted to 98.3% of all the development aid loans (OECF 1995: p. 8).

These statistical values seem to contradict the criticism often made that Japan substantially protected its private industry in 1990s through aid projects and still does today. However, the question remains to be answered whether invisible trade barriers and interest groups did not disadvantage foreign competitors when the orders for projects were distributed. Japan shares a preference of its own economic interests with other DAC countries; as long as the safeguarding of one's own interests is also contributing to effective development in the receiving countries, such a policy can be regarded as legitimate.

The aim of the so-called *Sector Program Loan* valuing JP¥ 20,844 billion was the careful strengthening of aid projects in important sectors like the educational and health systems, social welfare, transport, water resources and irrigation, in order to fight poverty and inequality in regional development (Kokusai kaihatsu jaanaru 1995: p. 137). The proportion of environmental development aid for this programme dedicated mainly to the infrastructure amounted to JP¥ 2,092 billion. This was mainly for constructing sewers for household and industrial waste and filter plants (INTEP 1995: p. 129); the environmental element amounted however to only 10% of the total credit.

Also, the *Rural Area Infrastructure Development Project* (Water Supply System), with a total budget of JP¥ 21 billion, helped the development of the infrastructure of villages and the main feature of this aid project was the construction of a water supply network. The OECF quotes the proportion of aid for the environment as valuing JP¥ 2,709 billion (OECF 1995: p. 110), although the environment aspect was only secondary as with the *Sector Program Loan*.

In contrast the loans as part of the *Bapedal Regional Monitoring Capacity Development Project* and of the *Denpasar Sewage Development Project* were primarily for the transfer of technology for the control and protection of the Indonesian environment and belonged without reservation to Japan's environmental aid programme.

The loan for the *Denpasar Sewerage Development Project* amounted JP¥ 5.4 billion. The aim of this project was the improvement of the public sewage system in the areas Denpasar, Sanur and Kuta on the island of Bali, the connection of private households to the sewage system and the purification of sewage by efficient filter plants. These measures had to contribute to an improvement in the water quality on the coast of Bali - an increasingly popular destination for tourists from all over the world (Kokusai kaihatsu jaanaru 1995: p. 136).

The *Environment Monitoring Improvement Project*, which was financed through a loan of JP¥ 2,935 billion, collected environmental data and improved the efficiency of the environmental laboratories of the health ministry, industry and local governments in Indonesia (OECF 1995: p. 143).

As described in Chapter 10.4.1 the OECF and the Japan Export Import Bank (JEXIM) were replaced by the Japan Bank for International Cooperation (JBIC).

The German *Kreditanstalt für Wiederaufbau* (KfW) is comparable to the OECF. The KfW is a government-owned bank for the development of the German economy, but at the same time providing financial support to governments, public enterprises and commercial banks engaged in microfinance and SME promotion in developing countries. Within the environmental technology sector, KfW was mainly active in the 1990s through the provision of loans to finance investments of German ET companies in developing countries and NICs and to finance the export and implementation of projects for installing and up-to-date environmental techniques. In 1996 38 environmental projects were developed under KfW's financial cooperation with a total volume of approximately EUR 320 million. Hence, similar to JICA and OECF, GTZ and KfW have been closely linked, though differing with Japan, a merging of the technical cooperation, grant aid and concessional loans is not yet on Germany's political agenda. The German government is actually concentrating on reform of the numerous implementing agencies of Germany's ODA, which includes the GTZ as by far the largest.

13.5 Environmental Technology Transfer by Private Industry

With the rapid increase in the expenditure of local administrations and of the government in Tokyo as a result of public pressure to find effective solutions for environmental problems in the early-1970s, private industry began to re-orientate itself more according to the increasing public interest in the protection of the environment. The companies for environmental technology differed in this respect from the automobile and high-tech industries, which acted almost exclusively according to purely economic considerations and which only gradually discovered the environment sector for their sales strategy (Park 1995: p. 4).

The Ebara Corporation was one of the leading Japanese companies in the environmental technology sector. Ebara was able to increase its turnover in the first half of the 1990s by 43% as a result of a growing inland demand for ET. In fiscal year 1995 the total turnover amounted to US$ 2 billion. A key element in Ebara's future business strategy was the expansion of its international operations. For this reason this company set up over 30 offices and subsidiaries in 16 countries - six of them in Asian countries such as Indonesia. Despite this its exports had a share of only 10% in the total turnover; planning strategists did not reckon in the short-term with a noticeable increase in export quotas to NICs and developing countries.

Other Japanese private companies for environmental technology like Kankyō Corporation (Yokohama) were making a change in their strategy in preparation for a growing market in environmental technology in the 1990s (JEC 1994: p. 102ff). Their limited engagement in NICs like Indonesia, however, provided them an option on the enormous demand for environmental technology which was expected in the mid-term and prepared them for the market requirements in that country. This was also one key-reason why German actors from the private sector started in the mid-1990s to strengthen their efforts in the transfer of German environmental technologies, preparing for the growing markets and enormous export potentials.

For Japan and Germany in the 1990s there did not exist a detailed overview of companies active in the environmental technology transfer sector. The number of German companies ranged somewhere between 2,500 and 8,000 (Halstrick-Schwenk 1994: p. 149; Wackerbauer 1995: p. 8). In contrast to Japan the quota of specialized SME active in the environmental technology sector was rather high but similar to Japan the number of large companies was also high, most of them offering a number of various environmental techniques such as Ebara and Kankyō Corporation. Today, the growing competition on the worldwide ET market leads to increased concentration. This is why the German Federation of Industrial Research Associations (AiF) is supporting the promotion of environmental technologies of SMEs.

13.6 Environmental Technology Transfer by Non-Governmental Organizations

In the 1990s in Japan, NGOs (*hi seifu soshiki*) – in contrast to the United States and most European nations – were still of low significance for development aid. This applied not only to the implementation of projects but also for taking influence towards reforms domestic development aid policy. While the Japanese government increased its subsidies for NGO projects from JP¥ 188.7 million in 1990 to JP¥ 449.8 million in fiscal year 1994 (MOFA 1995a: p. 226), Japan was with its "expenditure per head" of US$ 2.3 for NGOs far behind Germany (US$12.3), the United States (US$ 9.9) and Norway and Switzerland both with US$ 31, respectively. For a long time Japan's politicians and government officials regarded the activities of the non-governmental organizations as destructive and parasitic for the work of the government. It was only in 1989 that the contributions of NGOs to development aid was welcomed by the Japanese government for the first time; their acceptance started to exponentially increase throughout the 1990s (Hirabayashi 1994: p. 27). Nevertheless, the NGOs had to struggle with great financial and bureaucratic problems in the 1990s, which prevented them from being granted exemption from taxes and unequivocal legal status (OECD 1996: p. 36).

Whilst Akiko *Tsuru* and Keiko *Kuniyuki* declared great interest of their organizations "Japan NGO Network for Indonesia (JANNI)" and "Friends of the Earth" for the transfer of environmental technology to Indonesia, during an interview in 1995, they asserted that they do not have enough financial and human capacity to pursue this objective. The same was confirmed by the "International NGO Forum on Indonesian Development (INFID)", which was critical of the environmental compatibility of some Japanese ODA projects and inspired a discussion about Japan's development aid for the environment (Moriguchi 1995: p. 5).

In contrast to Japan NGOs have been playing a key-role in Germany's development aid for a long time. The activities of charitable and church organizations such as e.g. Misereor[138] were and are indispensable for Germany's development aid. Their contribution to environmental technology transfer was mainly in urban development and ecological agriculture and also in the training of the local people. Other NGOs such as e.g. WWF Germany and NABU[139] were

---

138 The German Catholic Bishops' Organization for Development Cooperation (see http://www.misereor.org/about-us.html – 04 October 2010).

139 The Nature and Biodiversity Conservation Union (NABU) is one of the oldest and largest environment associations in Germany (see – http://www.nabu.de/en/nabu – 04 October 2010).

mainly advising for a sustainable usage of the natural resources and training for environmental sound management (UBA 1999: p. 169).

The examination of the implementation of Japan's projects for the transfer of environmental technology as part of Japan's transnational environmental policy through development cooperation with Indonesia illustrated a clear interdependence between the three executing groups –the private economy, the local players and the central state agencies. In contrast to Germany, NGOs played a marginal role in Japan in the 1990s. In Japan some of the regions affected most during the phase of ecological ignorance in the 1950s/60s such as Yokkaichi and Kitakyushu, developed a key-role in transferring their experiences coping with pollution in close interaction with the local industry and regional and national governments. There are no such known outstanding examples for Germany's activities, not only illustrating an environmental and economic interest, but also authenticity that besides certain foreign, domestic, environmental and economic interests something what they have gone through should simply not happen again anywhere in the world.

In Japan there has clearly been a concentration on:

a) Projects for the development of human resources through opportunities for specific training,
b) Measures for the monitoring of environmental conditions, in order to take effective environment protection measures,
c) The provision of information on available environmental technologies.

This is an indication of a long-term planning of activities in the field of environmental technology transfer to NICs along the paradigm sustainability, whereby a certain trading interest should not be unrecognized. But this again might also be a driver for the necessary progress in our present system.

14 Evaluation

In the fifth part of the policy-cycle model, the evaluation is made, i.e. the measuring and appreciation of the political intentions which have been put into practice. Purposeful and effective actions brought about results which can be divided into three different categories:

- Instrumental forms of political action, which include environment conventions, laws and regulations,
- Practical political results, e.g. employment of environment experts and the change of environmental protection strategy, and
- Results of action, which can be measured in the case of NICs such as Indonesia with a change in the environmental quality

However, the evaluation can be carried out in very different ways, since it can be based on various standards and criteria (Schubert 1991: p. 75f). Thus, the assessment can be made in the light of e.g. foreign policy attempting to find out whether Japan has succeeded in giving its foreign policy more of an outline. Following the paradigm of sustainability, the following evaluation tries to assess the whole implementation process, using certain indicators like the extent and speed of transnational environmental policies through the transfer of ET to NICs. The broad effects – the "outcome" – of state activities on all parties and all those involved by the policy serve as the benchmark (Jann 1981: p. 26). From this evaluation a hierarchy of motives, values and interests can be deduced, also illustrating the powers of those involved, which have caused Japan to transfer environmental technology to NICs as part of its transnational environmental policy also through development aid.

Nevertheless, it must be recognized that any framework, by itself, is an imperfect tool for organizing and expressing the complexities and interrelationships encompassed by sustainable development. Ultimately, the choice of a framework and a core set of indicators and criteria must meet the needs and priorities of users, in this case the policy analysts, and use of indicators to monitor progress towards sustainable development. The Commission for Sustainable Development (CSD) stressed that any country wishing to use indicators in any systematic way, must develop its own programme drawing on the resources currently available (CSD 2001).

14.1 Criteria of Effects

Transnational environmental policies differ in their effectiveness. This can be established by measuring their depth, breadth, speed and exactness. Also, when judging environmental activity the economic costs should be taken into consideration, so that a cost-effect-analysis can be made.

14.1.1 Depth of Effect

Deterioration of the environment manifests itself in damage to the environment and/or to health of the people. This damage is the direct result of changes in the quality of the environment, e.g. the pollution of the air, which is detrimental to humans, animals and plants and is caused by the emission of pollutants produced by the exploitation of natural resources like brown coal/lignite. Factors like the population density, economic size and consumption of resources influence the quality of the environment and under certain circumstances can cause damage to the environment and to the health of the people.

According to a rising gradation model of environmental burdens taking into consideration the size of population, the size of the economy, the consumption of resources, the quality of the environment and the damage to the environment and to health, statements can be made about the depth of effect of activity for the sake of the environment. The depth of effect is greater the lower the measures taken begin on the gradation model.

Considering an annual growth rate of the Indonesian population by 2.24% and a degree of urbanization of 50% until 2020 as calculated in the 1990s, a forecasted increase in industrial production in Indonesia by the factor of thirteen by the year 2020 compared with 1970 and the resulting enormous demand for energy, serious damage to the environment can be expected as long as suitable countermeasures are not taken in time. Compensatory and restorative measures, which are applied to damage already occurred, have an extremely small depth of effect.

On the other hand initiatives for a more efficient use of resources through new technology or even towards a sustainable society have a greater depth of effect and are more effective, though take considerable longer to be realized.

14.1.2 Breadth of Effect

Transnational environmental policies through environmental technology transfer have spatial, temporal and factual results. The spatial breadth of effect ranges from locally limited measures to those in all areas of the environment. In time this ranges from one moment in time to many years and decades or even forever.

Hence, the breadth of effect can vary from the effective development of a small ecological system to all forms of life. The breadth of transnational environmental policies depends on its construction; i.e. selective, regionally limited measures stand beside forms of action with a universal effect. Depth of effect and breadth of effect correspond with each other. Restorative measures for the removal of an environmental problem, e.g. in Palembang, aim at an improvement of the environmental quality in this region and as such are limited in their breadth of effect. The deeper the environmental action goes into existing structures, the greater in general is the breadth of effect.

14.1.3 Speed of Effect

With the conversion of environmental development aid into practical measures, the individual elements of development cooperation are changed. For instance ICETT begins with training environment experts and the time when the training shows results in a NIC such as in the Indonesian environment policy lies a period of time which is called the speed of effect. This corresponds with the depth and breadth of effect: action with a low depth and a small breadth generally takes effect sooner. This is because effective changes have to go through numerous stages corresponding to the model of the policy-cycle before the quality of the environment is improved. The speed of effect depends on those actively involved (Prittwitz 1990: p. 60f). The smaller the delays the greater is the speed of effect.

14.1.4 Exactness of Effect

The fourth criterion for judging transnational environmental policies through the transfer of ET is the exactness of effect. This has a qualitative and a quantitative dimension. The former determines the precision with which an environmental problem and the factors which cause it - have been treated. In correspondence with the breadth, depth and speed of effect, the greatest exactness depends on the depth and extent of the procedure. Contrastingly, the quantitative exactness of effect measures how much influence has been exerted on environmental problems. It is in not dependent on the other criteria.

14.2 Criteria of Effects in Context

The training of Japanese state environment experts by ICETT, KITA, EMC and other institutions was arranged on a very broad basis (Natori 1993: p. 66). However, in addition to the language problems which arose during the training of

Indonesian civil servants as a result of their poor English[140], the limited training capacity had a vital influence on the speed of effect (EA, OECC and BAPEDAL 1994: p. 27f).

Partnerships between local and national organizations like JICA and OECF helped bringing together local experience, human capabilities and institutions for the formulation of effective aid. Important measures for environmental protection had been taken on the initiative of local governments and not of the central government in Tokyo, so that local experience played a vital role in Japan's implementation of the environmental technology transfer to NICs. Local organizations like ICETT and KITA, which were founded with the help of local and national governments and private industry, have a great potential for transferring the experience of the region in dealing with environmental problems to the developing nations and the experience of private industry in the development and use of environmental technology through the national development aid by means of, for instance, JICA. It can be expected that these local experiences are also partly embedded in ET developed by German companies, especially SMEs. But they do not yet receive a similar support as in Japan through prominent local initiatives such as ICETT and KITA.

It was proposed that the EMC had to help industrial firms to develop effective environmental technology for the Indonesian market. Here it was the job of the local and national institutions as well as the supervising ministries, e.g. the former MITI to develop contacts with private industry. But in reality the differing interests of the triad of local administrations, the central government and private companies hindered the coordination of their activities. In addition to a vision, differing from on-going activities, conflicts of interests between the key-partners and members involved in the development of the German ITUT also contributed to ITUT's liquidation.

Moreover, the author of this study discovered that within an aid organization like JICA, there was a substantial lack of internal communication between different departments. Interviews and conversations were always used by JICA employees in order to receive information about the work of another department, which possibly was also engaged with the transfer of special environmental technology to NICs such as Indonesia.[141] This was the case in Germany, where contacted experts always attempted to get a better overview of environmental technology transfer related activities within Germany, pointing to a need of information none of the established projects could satisfy .

---

140 This was observed by the author during a visit of ICETT in July 1995.
141 As observed by the author during numerous meetings and appointments with employees of JICA, IFIC, ECFA etc.

Because the local Japanese administrations, the government and private industry all depended on each other, the transfer of environmental technology achieved a great breadth and depth of effect, because the cooperation of this triad – though not unproblematic offered a vast range of possibilities. However, the lack of an exchange of information between individual departments in the national aid organizations and conflicts of interest and competence among institutions entrusted with transnational environmental policy and development, aid tasks impeded the best use of these criteria of effect and had an influence on the qualitative exactness of effect. In the German case these latent quarrels accumulated in the termination of ITUT's work.

Cooperation with private industry was and might still also be a guarantee for the comprehensive, effective training of civil servants from NICs such as Indonesia. Some time passed before trainees were able to make use of their know-how in different areas of environmental technology, which they gained in extensive courses, in their home countries, since they had to surmount numerous bureaucratic and institutional hurdles. However, a low speed of effect has in general a positive influence on effective development. Trainings were also on the project list of e.g. the German GTZ, but did not succeed in a successful, far-reaching integration of the German environmental technology enterprises, not only owning these technologies but also sharing their applied experiences through their respective development and implementation. This is also why GTZ projects were regarded critically as mainly consultancy activities, not necessarily bringing along the sustainable effects.

The numbers of representatives from NICs and developing countries trained through e.g. KITA and ICETT, do not give a clear indication about the breadth and depth of these training tools. But as most trainings are designed to train multiplicators according to "Train the Trainers", a rather sustainable effect can be assumed. Staying in contact through a successful alumni work can also be considered important for this.

A further crucial part of Japanese transnational environmental policy by a transfer of environmental technology is the improvement of environment monitoring. From the perspective of environment policy, this measure has a great breadth and depth of effect as it is the basis for all environmental action and hence sustainable development because the quality of the environment is subject to a detailed analysis regarding polluters etc. (Böhret 1990: p. 254). Through the monitoring work of the EMC, of other laboratories in Indonesia and of ICETT, KITA and IETC, elementary data and information were gathered, which laid groundwork for the development of an efficient transnational environmental policy through environmental technology transfer. In addition these Japanese institutions for development cooperation with their catalogues of concrete measures initiated a conversion of the information gathered into concrete policies.

Also, further environmental development projects, e.g. the *Integrated Air Quality Management for Jakarta* by JICA or the *Denpasar Development Project* of the OECF, were developed along the paradigm of sustainable development. While they were regionally limited, they had a long-term positive influence on the air quality in Jakarta and the water management in the Denpasar area.

The period of time between the start of all the projects mentioned and the moment when they reach their optimum effect, i.e. their speed of effect was usually smaller the faster the measures were put into effect. In these cases mainly end-of-pipe or cleansing technologies were applied. Delays resulted from constellations of interest, positions of power, legal conditions, electoral periods or political organization structures. These different components were taken into consideration during the planning of the projects so that only slight delays were likely. Nevertheless, an instable political climate in Indonesia at the end of the regency of President *Suharto* in 1998 and the financial and economic crisis severely hitting Indonesia and other Asian nations hampered the application of the development projects.

The exactness of effect of Japanese environmental aid through environmental technology transfer to NICs such as Indonesia is impossible to judge without a detailed impact analysis in the recipient country; this would go beyond the limits of this study. Even in the recipient countries it will be hard to differentiate which donor activity resulted in which effect, because the interaction of all measures in one country results in the quantifiable and qualifiable effects. Moreover, the evaluation of project activities of the Japanese and the German development aid organizations are usually performed by in-house staff; a weak point often criticized, not ensuring the necessary objectivity because of a lacking distance to the implementing agencies. There were no official evaluation results published on both Japan's and Germany's transnational environmental policies through environmental technology transfer to NICs in the 1990s.

14.3 Hierarchy of Motives

The implementation of different projects and their selection have shown that environment and development motives stood in the forefront of Japan's environmental technology transfer to NICs in the 1990s. Trading and therefore economic interests became more obvious in the case of Germany, though not necessarily questioning the intended sustainability effect, but supporting the environment and development interests through economic trading measures.

It would be short-sighted to claim that Japan's policy was merely a mean to gain a greater reputation in a new world order. Japanese politicians were conscious that their actions were in line with their claim of playing a leading role in the global environmental arena and closely watched from outside (Hashimoto 1996: p. 72). This explains the fact that the author's investigations concerning Japan's transnational environment policy were substantially supported by a lot of governmental institutions. The growing public transparency of the Japanese environment and development policies contributed to a sharpening of the profile of Japan's transnational environmental policy and contradicts old prejudices that interpreted all foreign political activity as purely economic driven.

Japan's strengthened position through assuming more responsibility had also consequences in internal politics, so that the Prime Minister Hashimoto made global environment policy an essential part of his government's work. In the eyes of many Japanese politicians, it was easier to gain international and re-gain national reputation through transnational environment engagement than through the so-called Peace Keeping Operations (PKO). They felt it also contributed to compensate domestic difficulties through e.g. corruption and a tense economic situation.

With the successful involvement of national and local governments and the private industries in the transfer of environmental technology, Japanese enterprises were situating themselves into a better starting position in the growing market for environmental technology in NICs, which was expected in the mid-term, also in comparison with e.g. Germany. Thus, export interests cannot be denied, even though in the short-term Japanese companies reckoned with more growth in demand for environmental technology in the (post-)industrialized nations such as USA, Germany and Japan. In the 1990s 90% of world trade in environmental technology was concentrated in these countries. Moreover, the share of NICs in Asia and Eastern Europe seemed rather insignificant, while representing in the medium and long-term an enormous potential for growth (Barnett 1993b: p. 11).

Today, the global market for environment technologies regarding energy efficiency, water management, mobility, environmentally sound power generation, material efficiency, waste management and recycling accounts for approximately EUR 1,400 billion according to a recent study by Roland Berger Strategy Consultants (BMU 2009: p. 11). Unfortunately, this study, as with many others in this area does not provide any information about its sources or way of estimation. Nevertheless, the USA, Germany and Japan are highly competitive markets, where large high-tech companies compete with each other. Hence, the availability of home-grown ET makes it difficult for foreign environmental technology companies to penetrate these markets. In consequence the focus of transnational trade in the environmental technology sector has shifted to NICs such as Brazil, China, India, with Indonesia falling only in the second tier. Today, Indonesia is not yet a major player in the world market for environmental technology, neither on the demand nor the supply side; however, the Indonesian government still announced highly ambitious emissions reduction targets. Indonesia is a country which is especially vulnerable to climate change, with severe impacts of a changing climate already being felt in various parts of its 17,000 islands. Indonesia's largest emissions originate from deforestation, land degradation and conversion.

In the 1990s all loans for environmental development cooperation with Indonesia were made without strings attached, i.e. Japanese companies were not necessarily assigned with the execution of a project. A data bank for environmental technology of the IETC was expected to help make it possible for recipient nations to select the most suitable from a large number of firms in different nations. At first the personal contacts with Japanese industry of the trainees on the training courses were of importance. However, the market for environmental technology has been influenced by increasing transnationalization from the late 1990s, so that the Indonesian side soon decided to examine tenders according to the principle of cost/effect. In consequence personal contacts produced only a limited export advantage for Japanese industry.

Securing sources of raw materials seemed to play only a subordinate role in Japan's environmental technology transfer to countries such as Indonesia, since none of the analyzed projects in the 1990s was primarily or secondarily concerned with natural resources. Critics will say that Japan wanted to continue exploiting Indonesian resources like the tropical rain forest. However, this would have been detrimental to Japan's reputation abroad and to the credibility of Japanese politicians, who wanted to assume a leading position in global environment politics. In the face of a growing environmental consciousness in Japan, this attitude could have costed votes in the next elections, resulting in a loss of power. Today, Japan is a strong advocate of the 3R Initiative, which aims to build a sound-material-cycle society through the effective use of resources and materials. Again many nations are critical of Japan's approach of being too economically driven, though the 3R Initiative was agreed upon at the G8 Sea Island Summit in the United States in 2004. These critics fade out the long-time planning, which is part of Japan's governmental and industrial policies, which furthermore is rather uncommon among the majority of the other industrialized countries these days.

Security interests for the avoidance of conflicts in the Asian Pacific area were no longer visible after the implementation of projects and played only a marginal role in Japan's transnational environmental policy. In consequence the

14 Evaluation   173

hierarchy of motives for environmental technology transfer to NICs had development and environment policy interests at the very top in 1990s. Since private industry was intimately involved in the implementation of the projects, export motives played an important but not predominant role. The considerations of foreign and internal policy ranged below these in the hierarchy. A responsible foreign policy had influence on a stronger national reputation and the hierarchy ended with the motives of security policy.

As shown sustainable development had and continues to have a large impact in the implementation of transnational environmental policies through environmental technology transfer to NICs by both Japan and Germany. Referring to the recurrent critics of Japan's rather economically orientated policies, the economic dimension is one of the three dimensions of sustainable development. The environmental technology transfer to NICs illustrates that trading aspects constituted a more important role for Germany than for Japan

The implementation of some ET transfer projects to Indonesia showed the close interaction of local, regional and national governments and (semi-) governmental organizations with private industry, whereas non-governmental actors played a secondary role in Japan in comparison with Germany. Typically, this interaction is evident in a successful implementation of policies in a sector where the nation-state is not the owner of the respective techniques, but has the role of pushing, paving the paths and coordinating national interests towards global sustainability. It is exactly the successful interaction of the key-actors in Japanese projects through which Japan as a nation-state gained reputation in international circles, making it a key-player in environmental technology transfer activities. This means that Dehnhard's theory was verified, despite a lacking engagement of civil society and at the face of necessary ODA reforms in Japan in the early 2000s. Today, Japan is on its way to improve the short-comings through its increased support of NGOs and by improving the quality of its development aid. Japan's unfortunate role in the discussions of the Whaling Convention and its lacking leadership in the international climate talks are examples contradicting this general judgement.

In comparison with Japan, Germany would substantially benefit from the establishment of a successful cooperation of all relevant actors' groups including the private sector also in its development cooperation activities. Through this, the German government would gain a clearer role and be in the position to also achieve deliverables ascribed to its active support, whereas it is mainly an intercessor for German enterprises. This position neither strengthens nor weakens Germany's influence in international circles, because its intensive competence in the environmental technology sector is highly visible, very much similar to Japan. Nevertheless, there is room to gain more power as a nation-state in transnational environmental policies with a successful cooperation of local and national governments, (semi)-governmental organizations and the private sector in the environmental technology transfer activities towards NICs.

15 Termination

Taking into account the control of the effects in the evaluation, the decision is made in the last section of the policy-cycle model, the termination phase, whether and how a programme is to be continued. In so far as ineffective or dysfunctional elements have appeared, an improvement by programmatic or implementative reactions become necessary. Thus, termination has a similar character as selection. If political action was successful or the framework conditions change to such an extent that the measures seem pointless, a programme can be discontinued completely (Schubert 1991: p. 76f). In general during termination a successful pattern of action is reinforced or a less successful one is changed (Beyme 1985: p. 12).

In the late 1990s Japanese authorities had not yet conducted studies to examine the effectiveness of their environmental projects through technology transfer. Rather, institutions were occupied in detail with the conceptual and organizational arrangement of different projects. At the same time changes were being made in order to improve the effectivity of individual elements; these changes included enlarging the staff of permanently employed and/or advisory environment experts in JICA and OECF, an increased budgeting of environment aid projects by the government and a strengthened cooperation with private industry.

The establishment of Environmental Management Centres in China and Thailand displayed that Tokyo regarded this form of environmental aid as elementary. The annual increase in the budget for Green Aid in the 1990s proved that there was no indication for a change in Japan's environmental aid through environmental technology because of a lack of success. Rather, one reckoned with an extension of Japan's programmes and the initiation of further projects.

Today, MOFA and others no longer explicitly report on their progress in environmental issues as part of ODA. Environmentally related activities are now embedded in the following twelve categories:

1. Aid Coordination
2. Cultural Grant Assistance
3. Debt Problems
4. Democratization
5. Disaster Reduction
6. Education
7. Gender & Development
8. Health
9. Human Security

10. International Digital Divide
11. MDGs
12. Water and Sanitation

In comparison to Japan, Germany decided not to continue the support of the ITUT, finally leading to the termination of its operation. Since then, the work of ITUT has not been integrated into any other institute. Hence, Germany is actually lacking an institute or initiative building a bridge between public and private initiatives in transferring environmental technologies to NICs. Also, training activities are only latently dealt with in development cooperation agencies such as GTZ and Inwent, recently being merged with DED to form the GIZ[142]. In consequence long-time there was no one specific actor assigned with these capacity building duties and hence in the position to effectively build up such an institute specialized in ET trainings. Obviously, German politics and the private sector assess the present efforts to transfer ET into NICs mainly by the private sector as sufficient.

---

142 Inwent - Capacity Building International, Germany, is a non-profit organization with worldwide operations dedicated to human resource development, advanced training and dialogue. Inwent is commissioned by the German Federal Government, the German business sector and the German Länder. (See http://www.inwent.org – 5 October 2010).

Part D – Conclusions

16 Conclusions

This study was on Japan's transnational environmental policy through environmental technology transfer to Newly Industrialized Countries (NICs) in the 1990s. The overriding research question addressed Japan's ability to solve the growing environmental problems in the rapidly industrializing countries through the transfer of appropriate environmental technologies and therefore, to take up the challenges for nation-states in this transnational sphere. A comparative element with Germany was applied to gain additional knowledge in answering the study's key questions and illustrate certain transferability to other nation-states.

The Modernization Theory, International Dependence Theory and Development Theory, Real Existing Socialism and Flying Geese Pattern of Development are among those theories, which are regarded as failed and hence, inappropriate to serve as a paradigm for this study. But the normative proposal of sustainable development claiming that future development cannot follow the model of the past and that ways are needed to achieve economic, social and ecological objectives at the same time, also considering long-term implications of decisions, has become the key reference term for future development and policy-making. Interestingly, the original meanings of sustainable development differ in Japan and Germany, though this study could not identify any indicators for substantially different understandings of both nations through their respective transnational environmental policies.

The present era of transformations is influencing the national policy formulation processes. However, for this development literature assumes a loose and overstretched notion of the term "globalization" with quantitatively and qualitatively different phenomena in environmental, economical and social relations. Interstate interactions involving non-governmental actors – individuals or organizations – are considered transnational from a political-scientific point of view. Moreover, based on the growing importance of non-governmental actors in the field of international environmental policies and development policies, especially in the case of environmental technology transfer through public-private-partnerships, it appeared justified to speak from an ongoing "transnationalization". This study could not identify the evidences for governments' constant loss of control and nation-states' ability to govern through this transnationalization and hence to hold discourse on "denationalization".

The meanings of "environment" and "technology" in politics and to the public are still rather unclear in Japan and Germany but also elsewhere. One reason is that "environment" can be structurally or functionally understood, whereas "technology" in contrast to "technique" implies the know-how required to develop and apply techniques and technical procedures. Today, novel machineries and techniques should no longer concentrate on production of goods and services in a most efficient and effective way, but also take into account potential negative and positive impacts on the environment. These new requirements are met by technologies where combining the terms environment and technology appeared appropriate, naming them "environmental technology". Of those international and national attempts to define "environmental technology", only the Japanese International Center for Environmental Technology Transfer (ICETT) was able to provide a comprehensive definition and description. For the others institutions in Japan, Germany but also on the international level, it appears like a strategic gambit of failing to define "environmental technology" for not limiting their activities.

"Environmental technology" can be separated into four categories: 1. Measuring technologies on the environment, 2. Cleansing technologies or end-of-pipe, 3. Cleaner technologies and 4. Clean or zero impact technologies. Even though only the installation of cleaner and clean technologies would meet the definition of sustainability by economic and ecological parameters, environmental technology has to take all four categories into account. This analysis guided the development of an operational definition of "environmental technology", which was thus far missing in scientific and political discussions.

Rapid urban and industrial growth has brought serious problems in NICs e.g. including air pollution from power and industrial plants and still other mobile sources such as vehicles, water pollution from depleting wastes, growing municipal and industrial wastes and the enormous extraction of resources. But NICs have also enjoyed certain advantages associated with being late such as the opportunity to learn from predecessors, reduced uncertainty about absolute and relative magnitudes of environmental risks and a rapid uptake of technological innovations. Moreover, the availability and transfer of a wide-range of low-cost environmental technology options can more closely match the country's environment.

The categorization of environmental problems into natural disasters, man-made environmental problems and those having immediate and exclusive consequences only on mankind through being the last link of the food chain demonstrated the challenge nation-states are confronted with in taking appropriate measures against them. Transnational environmental problems can have global, local/regional characters; hence, the policies developed by Japan or Germany in response to transnational environmental problems do not necessarily have lasting effects themselves. However, this study developed an operational definition of transnational environmental problems at the absence of such in the available literature. It demonstrated the necessity for transnational action but also provided the perspective for increased efficiency through mutually-sustained measures. This also made the dilemma clear: Modern states are confronted with a burden to make decisions but decisions made do not necessarily have an effect on the problem at the absence of transnational harmonized action. Nevertheless, the political system is given the task to master the new conditions at the sight of changes in the system bridging a separation of space and time, immersing national economies in a sea of global flows and forcing them to cope with an increasing number of environmental problems.

The analysis of the scientific discussions showed that social scientists differ in their assessment about the future of nation-states, politics and strategies capable of bearing policy formulation under this changing framework through transnationalization. Out of this four far reaching viewpoints were worked out, concluding that as long as modern industrial nations are organized as democracies with a parliamentary system, the state and its machinery should play the key-role in solving life-threatening problems for society, even if modified functions are appearing under the complex conditions of modern industrial societies. This study argued that globalization does not necessarily undermine the state, but includes transformations of state forms and policymaking. The transfer of environmental technology to NICs proceeds on the assumption of the existent, but probably reduced capacity of the nation-state to act. Consequently, this study paid special attention to the actors and how they were influencing Japan's transnational environmental policies. It also analyzed to what extent these actors were influenced from other non-national actors and which were relevant factors for policy formulation and the implementation based on Albrecht Dehnhard's neo-realistic theory that nation-states were gaining power in international circles.

The presented cases of the Whaling Convention and the Kyoto Protocol exhibited three different aspects relevant for this study:

1. The tendency through increasing transnationalization to prescribe certain sights, result in major conflicts. These conflicts occur out of the lacking consideration and scrutinizing of the different interests, so that they end up in opposition against a majority or a supposed hegemon.
2. The massive possibilities to influence the decision of governments from abroad, especially when they are not taking a clear position against a certain policy.
3. The environmental technology at the interface between economic interests and ecological considerations has a special role, which remained to be clarified in course of the policy-cycle analysis of this study.

The critical analysis and evaluation of Japan's transnational environmental policies in the 1990s required an investigation of the framework conditions within which these policies developed. The examples of imports of forest and agriculture products and imports of selected raw materials illustrated Japan's extensive dependence on the natural resources of other nation-states for its key-industry and national consumption. Their extraction and further processing came with substantial environmental impacts on the environment and development in NICs. In addition Japan's greenhouse gas emissions illustrated its large footprint on the Earth's ecosystem. In consequence Japan is largely contributing to the development of transnational and global environmental problems but has not unrestrictedly taken up the responsibility to lower or eliminate those problems; as depicted throughout this work, this is also due to economic interests.

Simultaneously, the analysis demonstrated that some interstate interactions may involve governments, but also other actors with industry, financial institutions and international and non-governmental actors playing a significant role. Additionally, it also reflected a key error in current globalization debates: The identification of the modern state with the nation-state.

The analysis along the policy-cycle model displayed that Japan's transnational environmental policy has been initiated and carried through in the regions and within different parts of society. Without the influence of the citizens' movement at the time of ecological ignorance of the governments, without the cooperation of local administrations and without private industry these policies, substantially involving development cooperation with exclusively federal means, would have been extremely limited. While the conflict of interest between different levels complicated the implementation, the consensus ultimately found was also to the advantage of the recipient nations.

The first step of the policy-cycle model, namely the initiation, is followed by the estimation, which brought along several changes in the 1990s. Thus, JICA opened a department for the environment, women in development and other global aspects, the Overseas Economic Cooperation Fund (OECF) – a department for environment and development, and both JICA and OECF were employing teams of experts to examine the effect of their development projects on the environment. The ODA Charta, the environment guidelines of the OECF and Agenda 21 of the United Nations Conference on Environment and Development (UNCED) formed the framework for Japan's development aid. Based on these conditions the third step of the policy-cycle model, the selection, the Basic Environment Law and the Basic Environment Plan were created. The Agenda 21 was put into effect with the help of a National Action Plan, which included the transfer of environmental technology as an important component. At the same time MITI promoted environmental cooperation with developing countries and NICs referencing the so-called Green Aid Plan. In the implementation the policies which were decided upon were converted into concrete projects. A triad of local organizations like ICETT, KITA and IETC, national organizations such as JICA and OECF as well as private industry players including consultant companies transferred mainly Japanese know-how about environmental technology and environment management in the form of training courses for civil servants in NICs such as Indonesia. In addition the transfer of environmental monitoring technology aided in tracing pollutants and the sources of pollution thereby contributing to an effective control of acute and nascent environmental problems as a result of swift economic growth.

Loans without strings attached and databanks for environmental technology, which interested actors were able to be accessed worldwide but also enabled Indonesian industry and the Indonesian government to select their contractual partners according to the principle of cost/effect. In spite of this Japan was chosen in many cases primarily as a result of Japan's substantial engagement in Southeast Asia.

The evaluation, as the fifth step in the policy-cycle model, determines the strength of various criteria a great breadth and depth of effect, so that Japan's transnational environmental policy through environmental technology transfer corresponds to the definition of sustainable development. However, no statements could be made about the exactness of effect of the projects as it would have required a detailed impact assessment in the recipient countries. In addition the Japanese side has not yet made an official evaluation study of current projects publicly available and relevant data cannot be found in the current development reports of the OECD or the World Bank. Also, the question about the exactness of effect of the policy adopted remains unanswered.

Similarly, this study could not describe the last step of the policy-cycle model, the termination, since an extensive evaluation of the policy adopted, which could lead to possible changes, has not yet taken place. However, the proportionate increase of environment experts at JICA and OECF and the regular increase in the budget of the Green Aid Plan indicated a desire for a quantitative and qualitative increase in environmental development cooperation, also through environmental technology transfer.

Under the proviso that Japan's environmental involvement proved efficient, synergetic effects were expected with the transfer of Japanese environmental technology. Thus, a sustainable development with Japanese support in NICs such as Indonesia could have served as a demonstration of competence; in this respect Japan would have ensured its potential for worldwide growth in the market for environmental technology. Nevertheless, Japan proved that it is striving to take more responsibility in a new world order through its commitment to the environment. Similarly, more intense environmental initiatives were partly responsible for improving the reputation of some Japanese politicians and gave them a better profile at home, at least in the short-term.

This study demonstrated that Japan's transnational environmental policy through environmental technology transfer to NICs such as Indonesia in the 1990s was led by numerous motives. These motives include environmental technology transfer as an element of Japan's transnational environmental policy on the political agenda. With the scant lucrative markets for Japanese environmental technology in Indonesia in the 1990s, the motivation for improving the environment dominated over the motivation of foreign policy. Certainly, aspects of improving the environment were also being addressed and put forward in order to gain a sound starting position for future markets in NICs such as Indonesia and increase its reputation both domestically and internationally. But in order to work out evidences for a trade-strategic action in Japan, a detailed analysis of private sector activities would be required.

The synthesis of the macro-analytic analysis - also through comparison with Germany - with the macro-analytic examinations of the project implementation made clear that environment and development interests ranked above those of foreign policy. Nevertheless, Japanese industry – with the support of the Japanese governmental organizations – created an ideal starting point for future market growth in environmental technology in NICs through being inextricably integrated in the transfer of environmental technology.

With reference to Albrecht Dehnhard's theory, this study exhibited that nation-states were the main actors in transnational environmental policies through environmental technology transfer to NICs. However, the national government had to closely interact with other key-actors such as local and regional governments, consultancies, the private sector and NGOs in order to guide its policy to success and hence obtain the desired effects. Contrary to Dehnhard's assumption it was not verified if Japan lost power and influence at the international level through its environmental technology transfer. This analysis illustrated the transformation of state forms, in which the government had to closely interact with other actors-groups; thus, the transfer of environmental technology to NICs proceeds on the reduced capacity of governments to act due to the influence of non-national actors, also impacting policy formulation and the implementation process. Due to the successful coordination of the various actors' interests and activities under a jointly defined policy, the modern nation-state gains a strong standing and subsequently more power in international circles which supports Dehnhard's neo-realistic theory. Nevertheless, this gain is destroyed by Japan's positions in other transnational environmental issues such as whaling and climate change, which are substantially criticized both internationally and domestically. In consequence one may deny Japan's leading role in transnational environmental policies these days, though it is holding such in the sector of environmental technology transfer to NICs.

17 Sources
17.1 Bibliography

Aachener Stiftung Kathy Beys (2004): Lexikon der Nachhaltigkeit (http://www. nachhaltigkeit.aachener-stiftung.de/2000/Definitionen.htm – 02 December 2004).

ADB (ed.) (1994): Financing Environmentally Sound Development, Manila.

ADB (1986): Environmental Planning and Management. Regional Symposium on Environmental and Natural Resources Planning, Manila.

Almeida, Celso (1993): Development and Transfer of Environmentally Sound Technologies in Manufacturing: A Survey, UNCTAD Discussion Papers, No. 58, Geneva.

Albrow, Martin (1998): Abschied vom Nationalstaat, Frankfurt/M., 1st Edition.

Altvater, Elmar & Birgit Mahnkopf (1997): Grenzen der Globalisierung. Ökonomie, Ökologie und Politik in der Weltgesellschaft, Münster, 3rd Edition.

Anter, Andreas (1998): Georg Jellineks wissenschaftliche Politik. Positionen, Kontexte, Wirkungslinien, in: Politische Vierteljahresschrift, September, p. 501-526.

Arase, David (1995): Buying Power. The Political Economy of Japan's Foreign Aid, London.

Arase, David (1994): Public-Private Sector Interest Co-ordination in Japan's ODA, in: Pacific Affairs, Vol. 67. p. 171-199.

Archer, Martin S. (1990): Preface, in: Albrow, M. and E. Kirg (eds.): Globalisation, Knowledge and Society, London, p.1-2.

Ashcroft, Brian (1998): Globalisation does have advantages, but it could also pose a greater threat to markets than is perceived, in: The Scotsman, Edinburgh, 24 August.

Auswärtiges Amt (1998): Außenpolitik für Umweltschutz (http://www.auswaertige-gesamt.de/www/de/infoservice/download/pdf/publikationen/auspolumw.pdf – 17 June 2006).

Ayukawa, Yurka & Yoichi Mizutani (2002): Japan…Could Do Better, Tokyo.

Bachrach, Peter & Moroton S. Baratz (1977): Macht und Armut. Eine theoretisch-empirische Untersuchung, New York, 1st Edition.

Bamyeh, Mohammed A. (1993): Transnationalism, in: Current Sociology, Vol.41, No. 3.

BAPEDAL (1994): Donor Projects Database, Jakarta.

Barbian, Thomas W. (1992a): Geschichte der Umweltpolitik, in: Dreyhaupt, Franz J. et al. (eds.): Umwelt Handwörterbuch. Umweltmanagement in der Praxis für Führungskräfte in Wirtschaft, Politik und Verwaltung, Berlin, p. 154-158.

Barbian, Thomas W. (1992b): Akteure und Programme der Umweltpolitik, in: Dreyhaupt et al. (eds.): Umwelt Handwörterbuch, Berlin, p. 159-162.

Barbier, Edward. B. et al. (1991): Technological Substitution Options for Controlling Greenhouse Gas Emission, in: Dornbusch, Rüdiger & James M. Poterba: Global Warming: Economic Policy Responses, Cambridge, p. 109-161.

Baring, Arnulf & Masamori Sase (eds.) (1977): Zwei zaghafte Riesen? Deutschland und Japan seit 1945, Stuttgart/Basel.

Barnett, Andrew (1993a): International Technology Transfer and Environmentally Sustainable Development, Paper prepared for the UNCTAD Workshop on the Transfer and Development of Environmentally Sound Technologies, Oslo.

Barnett, Andrew (1993b): Do Environmental Imperatives Present Novel Problems and Opportunities for the International Transfer of Technology? Paper prepared for the UNCTAD Workshop on the Transfer and Development of Environmentally Sound Technologies, Oslo.

Bebler, Anton (ed.) (1990): Contemporary political systems: classifications and typologies, London.

Beck, Ulrich (1998): Wie wird Demokratie im Zeitalter der Globalisierung möglich, in: Beck, Ulrich (ed.): Politik der Globalisierung, Frankfurt, p. 7-66.

Beck, Ulrich (1997): Weltrisikogesellschaft, Weltöffentlichkeit und globale Subpolitik, Wien.

Beck, Ulrich (1993): Die Erfindung des Politischen, Frankfurt/M.

Beck, Ulrich (1986): Risikogesellschaft – Auf dem Weg in eine andere Moderne, Frankfurt/M.

Beckmann, Armin (1984): Leben Wollen, Köln.

Berger, Peter (1987): The capitalist revolution – fifty propositions about prosperity, equality and liberty, Aldershot.

Berg-Schlosser, Dirk & Ferdinand Müller-Rommel (1997): Vergleichende Politikwissenschaft, Opladen, 3rd Edition.

Bersihand, Roger (1963): Geschichte Japans von den Anfängen bis zur Gegenwart, Stuttgart.

Beyme, Klaus von (1996): Theorie der Politik im Zeitalter der Transformation, in: Beyme, Klaus von & Claus Offe (eds.): Politische Theorien in der Ära der Transformation, Opladen, p. 9-29.

Beyme, Klaus von (1990): Die vergleichende Politikwissenchaft und der Paradigmenwechsel in der politischen Theorie, in: Politische Vierteljahresschrift, Issue 3, p. 457-474.
Beyme, Klaus von (1985): Policy Analysis und traditionelle Politikwissenchaft, in: Hartwich, Hans-Hermann (ed.): Policy-Forschung in der Bundesrepublik Deutschland. Ihr Selbstverständnis und ihr Verhältnis zu den Grundfragen der Politikwissenschaft, Opladen, p. 12.
BGB (Bundesgesetzblatt) II 1993.
BGB II 1991.
BGB II 1973.
BGB I 1990.
Bhagavan, M. R. (1990): The Technological Transformation of the Third World, Strategies & Prospects, London.
Biermann, Frank et al. (eds.) (2002): Global Environmental Change and the Nations State, Proceedings of the 2001 Berlin Conference on the Human Dimensions of Global Environmental Change, Potsdam.
Bizri, Omar F. (1992): Environmentally sound technologies: their status and prospects, in: United Nations (ed.): Environmentally Sound Technology for Sustainable Development, ATAS Bulletin, Issue 7, New York, p. 32-37.
BMU (2009): GreenTech made in Germany 2.0. Environmental Technology Atlas in Germany, München.
BMU (1994): Environmental Policy. German Strategy for Sustainable Development, Bonn.
BMZ (2010a): Approaches in bilateral cooperation between Germany and its partner countries (http://www.bmz.de/en/approaches/bilateral_developme nt_cooperation/approaches/ index.html – 13 July 2010).
BMZ (2010b): Entwicklung der deutschen ODA-Quote 1967-2008, Bonn (http://www.bmz.de/de/ministerium/haushalt/imDetail/Deutsche_ODA-Quote_1967-2008.pdf – 13 July 2010).
BMZ (1997): Entwicklungszusammenarbeit auf dem Prüfstand. Auswertung der 1994/1995 durchgeführten Evaluierung des BMZ, Bonn.
BMZ (1993): Konzept für die entwicklungspolitische Zusammenarbeit mit den Ländern Asiens, Bonn.
BMZ (1989): Technologien für Entwicklungsländer - Die Konkurrenz zwischen Gegenwart und Zukunft, Entwicklungspolitik Materialien Nr. 79, Bonn.
Böhret, Carl (1990): Folgen, Entwurf für eine aktive Politik gegen schleichende Katastrophen, Opladen.
Böhret, Carl et al. (1988): Innenpolitik und politische Theorie. Ein Studienbuch, Opladen, 3rd Edition.

Bolt, Bruce A. (1995): Erdbeben –Schlüssel zur Geodynamik, Heidelberg.

Bożyk, Pawel (2006). Newly Industrialized Countries, in: Globalization and the Transformation of Foreign Economic Policy.

Braun, Dietmar (1993): Zur Steuerbarkeit funktionaler Teilsysteme: Akteurtheoretische Sichtweisen funktionaler Differenzierung moderner Gesellschaften, in: Héritier, Adrienne (ed.): Policy Analyse. Kritik und Neuorientierung, Darmstadt, p. 199-222.

Brewer, Garry D. und P. de Leon (1983): The Foundations of Policy-Analysis, Homewood.

Brickwedde, Fritz (1995): Umweltentlastung durch Innovation, in: Bundesministerium für Umwelt, Naturschutz und Reaktorsicherheit (ed.): Umwelt, 7-8/1995, p. 261f.

Brücher, Wolfgang (1982): Industriegeographie, Braunschweig.

Brooks, Harvey (1992): The concept of sustainable development and environmentally sound technology, in: United Nations (ed.): Environmentally Sound Technology for Sustainable Development, ATAS Bulletin, Issue 7, New York, p. 19-24.

Brown Weiss, Edith, Daniel B. Magraw & Paul C. Szasz (1992): International environmental law: basic instruments and references, New York.

Buck, Eugene H. (1997): Whale Conservation and Whaling, in: Oceans & Coastal Resources: A Briefing Book, Congressional Research Service Report 97-588 ENR, Washington D.C.

BUND & Misereor (eds.) (1996): Zukunftsfähiges Deutschland. Ein Beitrag zu einer global nachhaltigen Entwicklung, Berlin.

Carson, Rachel (1965): Der stumme Frühling, Frankfurt.

Castells, Manuel (2010): The Rise of the Network Society, The Information Age: Economy, Society and Culture, 2nd Edition, Oxford.

Castells, Manuel (2004): The Power of Identity, The Information Age: Economy, Society and Culture, 2nd Edition, Malden.

Castells, Manuel (2000): End of Millennium, The Information Age: Economy, Society and Culture, 2nd Edition, Oxford.

Chen, Shaohua & Ravallion, Martin (2008): The developing world is poorer than we thought, but no less successful in the fight against poverty, Policy Research Working Paper Series 4703, The World Bank.

China Greentech Initiative (2009): The China Greentech Report 2009 – (http://www.china-greentech.com/report – 14 July 2010).

Chichilnisky, Graciela (1982): Basic needs and the North/South debate, New York.

Choucri, Nazli (ed.) (1993): Introduction: Theoretical, empirical and Policy Perspectives, in: Choucri, Nazli: Global Accord: environmental challenges and international responses, Cambridge/Massachusetts, p.1-41.

CIA (2010): The World Fact Book (https://www.cia.gov/library/ publications/the-world-factbook/rankorder/2186rank.html – 5 August 2010).

Clapp, Jennifer (1997): Threats to the Environment in an Era of Globalization: An End of State Sovereignty? In: Schrecker, Ted (ed): Surviving Globalism, The Social and Environmental Challenges, New York, p. 123-140.

Clark, Ian (1998): Beyond the Great Divide: Globalization and Theory of International Relations, in: Review of International Studies, 24, p. 479-498.

Colomy, Paul (1990): Revisions and Progress in Differential Theory, in: Alexander, Jeffrey C. & Paul Colomy (eds.): Differential Theory and Social Change, New York, p. 119-163.

Colomy, Paul (1985): Uneven Structural Differentiation. Toward a Comparative Approach, in: Alexander, Jeffrey C. (ed.): Neofunctionalism, Beverly Hills, p. 131-156.

Commission on Global Governance (Ed.) (1995): Our Global Neighbourhood. The Report of the Commission on Global Governance, Oxford.

Cramer, Jacqueline and W.C.L. Zegveld (1991): The Future Role of Technology in Environmental Management, in: Futures, June, p. 451-468.

Crozier, Michael et al. (1975): The Crisis of Democracy: Report on the Governability of Democracies to the Trilateral Commission, New York.

CSD (2001): Indicators of Sustainable Development, New York (http: http://www.un.org/esa/sustdev/publications/indisd-mg2001.pdf – 15 Sept. 2010).

DEG and Commission of the European Communities (ed.) (1993): Thailand – Environmental Technology Study, Cologne.

DEG (1996): Thailand – Umweltstudie – Update, Cologne.

DEG (1993): Analyse des Kooperationspotentials für deutsche und europäische Unternehmen im Bereich des industriellen Umweltschutzes in Indien, Cologne.

Dehnhard, Albrecht (1996): Dimensionen staatlichen Handelns. Staatstheorie in der Tradition Hermann Hellers, Tübingen.

Dettmer, Hans A. (1965): Grundzüge der Geschichte Japans, Darmstadt.

Deutscher Bundestag, 13. Wahlperiode (1998): Unterrichtung durch die Bundesregierung. Umweltbericht 1998. Bericht über die Umweltpolitik der 13. Legislaturperiode, Drucksache 13/10735.

Deutscher Bundestag, 13. Wahlperiode (1995): Unterrichtung durch die Bundesregierung. Zehnter Bericht zur Entwicklungspolitik der Bundesregierung, Drucksache 13/3342.
Deutsch-Japanischer Kooperationsrat für Hochtechnologie und Umwelttechnik (DJR) (1999): Arbeitsbericht 1998, Bonn.
DIE (1990): Zur Notwendigkeit einer Verstärkung der wissenschaflichen Kooperationspolitik, Entwicklungspolitik aktuell, 001, Bonn.
Dierke Weltatlas (2008): Deutschland – Raumordnung, Braunschweig.
Dolzer, Rudolf (1998): Global Environmental Issues: The Genuine Area of Globalization, in: Journal of Transnational Law & Policy, Vol. 7:2, Spring, p. 157-179.
Dryzek, John S. (1997): The Politics of the Earth. Environmental Discourses, New York.
Dye, Thomas R. (1976): Policy Analysis. What Governments Do, Why They Do It, and What Difference It Makes, Tuscaloosa/Alabama.

EA (2001) : State of the Global Environment at a Glance. Acid Deposition (http://www.eic.or.jp/eanet/topic/acid_situ.html – 17 July 2010).
EA (1995): Quality of the Environment in Japan, Tokyo.
EA (1988): White Paper on the Environment, Tokyo.
EA - Committee on Global Environmental Problems, Japan's Activities to cope with Global Environmental Problems (1988): Japan's Contribution Toward a Better Global Environment, Tokyo.
EA, Overseas Environmental Co-operation Centre (OECC) and BAPEDAL (1994): The Republic of Indonesia, The Study on the Promotion of Environmental Protection Technology Transfer, Executive Summary, Tokyo.
Easton, David (1965): A Framework for Political Analysis, New York.
Eblinghaus, Helga & Armin Stickler (1998): Nachhaltigkeit und Macht. Zur Kritik von Sustainable Development, Frankfurt.
Economic and political weekly (1994): Paradigma shift, Vol. 29, No. 4, p. 169.
Edelman, Murray (1976): Politik als Ritual. Die symbolische Funktion staatlicher Institutionen und politischen Handelns, Frankfurt/New York.
Enquete-Kommission "Schutz des Menschen und der Umwelt" des 13. Deutschen Bundestages (1998): Abschlußbericht. Konzept Nachhaltigkeit. Vom Leitbild zur Umsetzung, Bonn.
ENTRI (2010): Environmental Treaties and Resource Indicators, Columbia University (http://sedac.ciesin.org/entri – 14 February 2011).
EMC (1995): Functions & Activities, Jakarta.
EPA (1996): Air Quality Criteria for Particulate Matter, Washington.

European Commission (1997): Eurostat Jahrbuch '97, Europa im Blick der Statistik, Luxembourg, 3. Edition.
European Commission (Ed.) (1994): The Market for Environmental Protection Equipment and Services in Indonesia, Brussels.
European Environment Agency (2010): Environmental technology introduction (http://www.eea.europa.eu/themes/technology/environmental-technology-introduction – 14 July 2010).
Evans, Peter (1994): Japan's Green Aid, in: The China Business Review, p. 39-43.
EX Corporation (Ed.) (1992): Indonesia, Resources and the Environment, Toward an international Co-operation for the Environmental Management, Tokyo.

Fasbender, Karl (1995): Erhaltung des Regenwaldes und ländliche Migration: Zielkonflikte am Beispiel Indonesiens, in: Hein, Wolfgang (ed.): Umweltorientierte Entwicklungspolitik, Hamburg, 2. Edition, p. 361-376.
Flüchter, Winfried (1998): (1998) Japan Weltwirtschaftsmacht mit Raum- und Ressourcenproblemen. In: Geographie heute, No. 158, p. 2-7
Flüchter, Winfried (1994): Geographische Fragestellungen, Strukturen, Probleme, in: Mayer, Hansjürgen & Manfred Pohl (eds.): Länderbericht Japan, Bonn, p.17-53.
Flyvbjerg, Bent (2001): Making Social Science Matter. What social inquiry fails and how it can succeed again, Cambridge.
Förstner, Ulrich (1995): Umweltschutztechnik, Berlin.
Forrest, Richard A. (1991): Japanese Aid and the Environment, in: The Ecologist, Vol. 21, No. 1, p. 24-32.
Foundation for the Rights o Future Generations (SRZG) (2004): Definitionen von Nachhaltigkeit im deutschsprachigen Raum (http://www.srzg.de/ndeutsch/5publik/Nachhaltigkeitslexikon.html – 2 December 2004).
Franke, Konrad (1987): Von der "Sowjetischen Besatzzone" zum eigenen Staat: Die Geschichte der DDR, in Pleticha, Heinrich (ed.): Deutsche Geschichte, Volume 12, Geteiltes Deutschland Nach 1945, Gütersloh, p.239-286.

Geiß, Jan (2002): Akteure, Handlungsebenen und Problemlösungsansätze im System des Sustainable Development, in: Sebaldt, Martin (ed.): Sustainable Development – Utopie order realistische Vision? Karriere und Zukunft einer entwicklungspolitischen Strategie, Hamburg, p. 112-157.
Gerlach, Irene (1988): Politikwissenschaft: Methoden und Instrumente, in: Bellers, Jürgen & Rüdiger Robert (ed.): Politikwissenschaft 1– Grundkurs, Münster, p. 154-197.

German Advisory Council on Global Change (WBGU) (1996): Herausforderungen für die deutsche Wissenschaft, Jahresgutachten 1996, Berlin (http://www.wbgu.de/WBGU/wbgu_auftrag_en.html – 15 July 2004).

German Environment Agency & German Federal Ministry of Environment (eds.) (2007): Umweltpolitische Innovations- und Wachstumsmärkte aus Sicht der Unternehmen, Dessau/Berlin.

German Trade & Invest (2010): Branchenbarometer – Asien / Pazifik Umwelttechnik, June 2010.

Giddens, Anthony (1985): The Nation-State and Violence, Cambridge.

Goldblatt, David et al. (1997): Economic Globalisation and the Nation State: Shifting Balances of Power, in: Alternatives, Social Transformation and Humane Governance, Vol. 22, No. 3, July-Sept., p. 269-285.

Gottlieb, Robert (2001): Environmentalism Unbound: Exploring New Pathways for Change, Cambridge/Mass.

Government of Germany (2002): Perspectives for Germany. Our Strategy for Sustainable Development, Berlin (http://www.nachhaltigkeits-rat.de/service/download_e/pdf/Perspectives_for_Germany.pdf – 6 December 2004).

Government of Japan (2002): The 2002/2003 Research Plan for the Japanese Whale Research Program under Special Permit in the Antarctic (JARPA), February 2002.

Government of Japan (1995): Outline of the Basic Environment Plan, in: Japan Environment Summary, Vol. 22, No. 5, p. 1-4.

Government of Japan (1994): National Action Plan for Agenda 21, Tokyo.

Graßl, Hartmut (2007): Klimawandel: Was stimmt? Die wichtigsten Antworten, Freiburg.

Green, Carl J.(1994): Japan's Growing Leadership in Global Development, in: Southeast Asian Studies Review, Winter-Spring, p. 101 - 118.

Greenpeace (2002): http://www.greenpeace.org/ – 05 October 2010.

Greenpeace (2007): Financial Grants and Votes for Whaling, Japan's Fisheries Aid and Links with the St. Kitts Declaration, February 2007.

Grundmann, Reiner (1999): Transnationale Umweltpolitik zum Schutz der Ozonschicht: USA

GTZ (1997): Fachgespräch: Technologietransfer und Stärkung technologischer Kompetenz, 23 April 1997, Eschborn.

Guillén, Mauro (2003): Multinationals, Ideology, and Organized Labor, in: Guillén, Mauro (ed.): The Limits of Convergence, p. 123-156.

Guardian, The (2010): Tokyo kicks off carbon trading scheme Japanese metropolis launches Asia's first emissions cap-and-trade scheme, 8 April 2010 (http://www.guardian.co.uk/environment/2010/apr/08/tokyo-carbon-trading-scheme – 21 July 2010).

Haas, Peter M., Robert O. Keohane and Marc A. Levy (1993): Institutions for the earth: source of effective international environmental protection, Cambridge.
Haber, Wolfgang (1992): Umweltbegriff, in: Dreyhaupt, Franz Joseph et al. (eds.): Umwelt-Handwörterbuch: Umweltmanagement in der Praxis für Führungskräfte in Wirtschaft, Politik und Verwaltung, Berlin, p. 2-6.
Habermas, Jürgen (1998): Jenseits des Nationalstaats? Bemerkungen zu Folgeproblemen der wirtschaftlichen Globalisierung, in: Beck, Ulrich (ed.): Politik der Globalisierung, Frankfurt, p. 67-84.
Habermas, Jürgen (1985): Die Neue Unübersichtlichkeit. Kleine Politische Schriften V. Suhrkamp Verlag, Frankfurt/M. 1985.
Hague, Rod et al. (1998): Comparative Government and Politics: An Introduction, Houndsmill.
Halstrick-Schwenk, Marianne et al. (1994): Die umwelttechnische Industrie in der Bundesrepublik Deutschland, Untersuchungen des Rheinisch-Westfälischen Instituts für Wirtschaftsforschung, Nr. 12, Essen.
Hajer, Maarten A. (1992): The politics of environmental performance review: choice in design, in: Lykke, Erik (ed.): Achieving Environmental Goals, The Concept and Practice of Environmental Performance Review, London, p. 25-40.
Hajnal, Peter (1989): The Seven-Power Summit: Documents from the Summits of Industrialised Nations, 1975-1989, Millwood.
Hammann, Peter & Heiko Mittag (1986): The marketing of Industrial Technology Through Licensing, in: Backhaus, Klaus & David T. Wilson (eds.): Industrial marketing, Berlin, p. 224-242.
Hammitzsch, Horst (ed.) (1981): Japan-Handbuch, Wiesbaden.
Harben, P. (1999): The Industrial Minerals Handbook, Industrial Minerals Information, Surrey.
Harborth, Hans-Jürgen (1992): Sustainable Development – Nachhaltige Entwicklung, in: Nohlen, Dieter & Franz Nuscheler (eds.): Handbuch der Dritten Welt, Bonn.
Hartmann, Jürgen (1995): Vergleichende Politikwissenschaft. Ein Lehrbuch. Frankfurt.
Hartmann, Jürgen (1992): Politik in Japan. Das Innenleben einer Wirtschaftsweltmacht, Frankfurt.
Hashimoto, Kôhei (1995): Senryaku Enjo. Chûtô wahei shien to ODA no shôraizô. Tokyo.
Hashimoto, Ryûtarô (1997): Statement by Prime Minister of Japan at UNGASS, 23 June 1997.

Hashimoto, Ryûtarô (1996): Antrittsrede des Ministerpräsidenten am 22. Januar 1996 vor dem japanischen Parlament, in: Japan Echo, Nr. 1, p. 71-75.

Hauff, Michael von (1997): Der Markt für Umwelttechnik, WISU, Kaiserslautern.

Hawken, Paul et al. (2000): Ökokapitalismus. Die industrielle Revolution des 21. Jahrhunderts, Gütersloh.

Heinze, Thomas (1995): Qualitative Sozialforschung: Erfahrungen, Probleme und Perspektiven, Opladen.

Heller, Hermann (1992): Gesammelten Schriften, 2nd Edition, Tübingen.

Helmut Kaiser Consultancy (Ed.) (1997): The environmental technology markets in Asia. India, Indonesia and Thailand, Berlin.

Héritier, Adrienne (1993): Policy-Analyse. Elemente der Kritik und Perspektiven der Neuorientierung, in: Héritier, Adrienne (ed.): Policy-Analyse. Kritik und Neuorientierung, Politische Vierteljahresschrift, Sonderheft 24/ 93, Opladen, p. 9-36.

Herrera, Amilcar et al. (1976): Catastrophe or New Society, Ottawa.

Herz, John H. (1976): The Nation-State and the Crisis of World Politics, New York.

Hilllebrand, Wolfgang et al. (1994a): Stärkung technologischer Kompetenz in Entwicklungsländern, Berlin.

Hilllebrand, Wolfgang et al. (1994b): Strengthening Technological Capability in Developing Countries – Lessons from German Technical Cooperation, Berlin.

Hillebrand, Wolfgang (1993): Stärkung technologischer Kompetenz in Thailand, Berlin.

Hirabayashi, Hiroshi (1994): Changes in the International Environment & the Direction of Japan's ODA, in: Japan 21st, Vol. 39, p. 23-27.

Hirst, Paul (1993): Globalization is Fashionable but it is a Myth, in: The Guardian, 22 March.

Hirst, Paul & G. Thompson (1996): Globalisation in Question: The International Economy and the Possibilities of Governance, Oxford.

Hobbes, Thomas (1996): Leviathan, Hamburg.

Hohn, Hans-Willy & Uwe Schimank (1990): Konflikte und Gleichgewichte im Forschungssystem. Akteurskonstellationen und Entwicklungs-pfade in der staatlich finanzierten außeruniversiäten Forschung, Frankfurt/M.

Holliman, Jonathan (1990): Environmentalism with a Global Scope, in: Japan Quarterly, July-September 1990.

Honecker, Erich (1971): Bericht des Zentralkomitees an den VIII. Parteitag der Sozialistischen Einheitspartei Deutschlands, Berlin.

Hoshino, Yoshiro (1992): Japan's Post-Second World War Environmental Problems, in: Ui, Jun (ed.): Industrial Pollution in Japan, Tokyo, p. 64-76.

Hünemörder, Kai (2004): Die Frühgeschichte der globalen Umweltkrise und die Formierung der deutschen Umweltpolitik, Stuttgart.

Imamura, Tsunao (1989): Environmental Responsibilities at the National Level: The Environment Agency, in: Tsuru, Shigeto and Helmut Weidner (ed.): Environmental Policy in Japan, Berlin, p. 43-53.

Imura, Hidefumi & Miranda Schreurs (eds.) (2005a): Environmental Policy in Japan, Northampton.

Imura, Hidefumi, Ryota Shinohara and Koji Himi (2005b): Environmental Industries and Technologies in Japan, in: Imura, Hidefumi & Miranda Schreurs (eds.): Environmental Policy in Japan, Northampton, p. 266-284.

Imura, Hidefumi (1993): Air Pollution Control Policies and the Changing Attitudes of the Public and Industry: Paradigmatic Changes in Environmental Management in Japan, in: Paoletto, Glen (ed.): Environmental Pollution Control. The Japanese Experience. Papers presented at the UNU International Symposium on Eco-Restructuring, Tokyo 5-7 July 1993, p. 55-85.

International Centre of Environmental Technology Transfer (ICETT) (2001): http://www. icett.or.jp/ – 15 May 2010.

International Centre of Environmental Technology Transfer (ICETT) (1995): Palembang Eco-Phoenix Plan Project 1993-1994, Final Study Report, Yokkaichi.

International Centre of Environmental Technology Transfer (1994a): The History of Pollution and Environmental Restoration in Yokkaichi, For the Sake of the Global Environment, Yokkaichi.

International Centre of Environmental Technology Transfer (1994b) Palembang Eco-Phoenix Plan Project 1993, Outline of the Study Results, Yokkaichi.

International Environmental Planning Centre (INTEP) (1995): Indoneshia kokuni okeru shinyô seikatsu haisui toshi haikibutsu tekisei shori shisutemu no sentaku shuhô no kaihatsuni kansuru kenkyû, Tokyo.

International Institute for Sustainable Development et al. (2004): National Strategies and Initiatives for Sustainable Development: A 19-Country Analysis of Strategic and Co-ordinated Action, Winnipeg.

Internationales Transferzentrum für Umwelttechnik (2001): http://www.itut.de – 27 February 2005.

Internationales Transferzentrum für Umwelttechnik (ITUT) (1996): Clean technologies for a better environment, Leipzig.

International Whaling Commission (2010): Whale Population Estimates, http://iwcoffice.org/conservation/estimate.htm#table – 1 July 2010.

Internationales Wirtschaftsforum Regenerative Energien (IWR) (2010): CO_2 Ausstoß (http://www.iwr.de/klima/ausstoss_welt.html – 22 July 2010).

IPligence (2009): Breakdown by geographic IP Location & Population Density & Internet Distribution, (http://www.ipligence.com/worldmap/ – 8 December 2009).

Jachtenfuchs, Markus (1998): Entgrenzung und politische Steuerung (Kommentar), in: Kohler-Koch, Beater (eds.): Regieren in entgrenzten Räumen, Opladen, p. 235-245.

Jann, Werner (1983): Policy-Forschung – ein sinnvoller Schwerpunkt der Politikwissenschaft? in: Aus Politik und Zeitgeschichte, 47/1983, p. 26-38.

Jann, Werner (1981): Categories der Policy-Forschung. Speyer.

Jänicke, Martin (2009): Geschichte der deutschen Umweltpolitik, in: Bundeszentral für politische Bildung (http://www.bpb.de/themen/DG6UQ7,0,Ge schichte_der_deutschen_Umweltpolitik.html – 23 July 2010).

Jänicke, Martin (1993): Vom Staatsversagen zur politischen Modernisierung? Ein System aus Verlegenheitslösungen sucht seine Form, in: Böhret, Carl & Göttrik Wewer (Eds.): Regieren im 21. Jahrhundert – zwischen Globalisierung und Regionalisierung, Pladen, p. 63-77.

Jänicke, Martin (1988): Ökologische Modernisierung. Optionen und Restriktionen präventiver Umweltpolitik, in: Simonis, Udo E. (ed.): Präventive Umweltpolitik, Frankfurt/Main, p. 13-26.

Jänicke, Martin (1986): Staatsversagen. Die Ohnmacht der Politik in der Industriegesellschaft, München.

Jänicke, Martin (1979): Wie das Industriesystem von seinen Missständen profitiert, Opladen.

Japan Environment Corporation (JEC) (1994): Kankyō hozen gijutsu iten sokushin chôsa hôkokusho - Indoneshia kyôwako , Tokyo.

Japan International Co-operation Agency (2010): Annual Report 2009, Tokyo.

Japan International Co-operation Agency (1996): Annual Report 1996, Tokyo.

Japan International Co-operation Agency (1995a): Annual Report 1995, Tokyo.

Japan International Co-operation Agency (1995b): Environmental Impact Assessment. Implementation of Environmental Consideration in JICA, Tokyo.

Japan International Co-operation Agency (1995c): The Study on the Integrated Air Quality Management for Jakarta Metropolitan Area in the Republic of Indonesia, Inception Report, Tokyo.

Japan International Co-operation Agency (1994a): Annual Report 1994, Tokyo.

Japan International Co-operation Agency (1994b): Kokusai kyōryoku jiyōdan: Indoneshia - kunibetsu enjo kenkyūkai hōkokusho (dai 2ji), Tokyo.

Japan International Co-operation Agency (1994c): Kokusai kyôryoku jiyôdan: Indoneshia - kunibetsu enjo kenkyûkai (genjô bunseki shiryô), Tokyo.
Japan International Co-operation Agency (1994d): The Second Country Study for Japan's Official Development Assistance to the Republic of Indonesia, Tokyo.
Japan International Co-operation Agency (1993): Study on Appropriate Environmental Protection Measures for Developing Countries, Tokyo.
Japan International Co-operation Agency (1988): JICA Aid Study Group. "Environment": Sectoral Study for Development Assistance, Tokyo.
Japan Ministry of Public Management, Home Affairs, Posts and Telecommunications (1990): http://www.stat.go.jp/english/1.htm – 15 July 2004.
Jaspers, Karl (1949): Vom Ursprung und Ziel der Geschichte, München.
Jellinek, Georg (1922): Allgemeine Staatslehre (1900), 3rd Edition, 4th Reprint, Berlin.
Jha, Veena und Ana Paola Teixeira (1994): Are Environmentally Sound Technologies The Emperor's New Clothes? UNCTAD Discussion Papers, No. 89, Geneva.
Johnston, Eric (2009): Japan under fire for laying low in Copenhagen, in: The Japan Times, 12 Dec. 2009.
Jomo, K.S. (1997): Southeast Asia's Misunderstood Miracle. Industrial Policy and Economic Development in Thailand, Malaysia and Indonesia, Boulder.
Jomo, K.S. (2001): Growth After the Asian Crisis: What Remains of the East Asian Model? UNCTAD G-24 Discussion Paper Series, No. 10, March 2001.
Jones, R.J. Barry (1995): Globalisation and Interdependence in the International Political Economy: Rhetoric and Reality, London.

Kaase, Max & Kenneth Newton (1995): Beliefs in government, Oxford.
Kaiser, Karl (1969): Transnationale Politik, in: Czempiel, Ernst-Otto (ed.): Die anachronistische Souveränität, Opladen, p. 80-109.
Kaiser, Karl et al. (1991): Internationale Klimapolitik, Eine Zwischenbilanz und ein Vorschlag zum Abschluss der Klimakonvention, Bonn, p. 13.
Kato, Kozo (2002): The Web of Power, Japanese and German Development Cooperation Policy, Oxford.
Katô, Saburô (1999): Chugoku nk kankyō Jijyô to Taisaki Senryaku, in: *Katô*, Saburô (1998): Junkan Shokai Sôzô no Jyôken. Nihon ga ikinokoru tame no 14 shyô, Tokyo.
Kawaguchi, Yoriko (2001): Statement of the Minister of Environment of Japan, COP-6bis, Bonn, 23 July 2001.

Kaya, Yoichi & Keiichi Yokobori (eds.) (1997): Environment, Energy, and Economy. Strategies for Sustainability, Tokyo.
Keizai Koho Center (1999): Japan 2000. An International Comparison, Tokyo.
Kelley, Donald R. et al. (1976): The Economic Superpowers and the Environment, The United States, the Soviet Union and Japan, San Francisco.
Keohane, Robert O. & Joseph S. Nye (1977): Power and Interdependence: World Politics in Transition, Boston.
Keohane, Robert O. & Joseph S. Nye (1971): Introduction, in: Keohane, Robert O. & Joseph S. Nye (eds.): Transnational Relations and World Politics, Cambridge, p. xii-xvi.
Kevenhörster, Paul & Dirk van den Boom (2009): Entwicklungspolitik, Wiesbaden.
Kevenhörster, Paul (2006): Die Entwicklungspolitik Japans auf dem Prüfstand: Evaluationen und ihre Folgen, in: Japan aktuell 3/2006, p. 37-53.
Kevenhörster, Paul (1995): Japan als internationaler Akteur: das Instrument der multilateralen Entwicklungshilfe, in: Asien, 01. Oktober 1995, p. 5-21.
Kevenhörster, Paul (1993a): Japan: die führende Wirtschaftsmacht auf der Suche nach einer neuen Rolle in der Weltpolitik, in: Konrad-Adenauer-Stiftung (ed.): Auslandsinformationen, St. Augustin, October 1993, p. 14-19.
Kevenhörster, Paul (1993b): Japan, Außenpolitik im Aufbruch, Opladen.
KITA (2009): Number of Overseas Participants received by KITA – (http://www.kita.or.jp/english/e_kensyu_jisseki.html – 01 Sept. 2010).
Klitgaard, Robert (1994): Taking Culture into Account, in: Serageldin, Ismail et al. (eds.): Culture and Development in Africa : Proceedings of an International Conference Held at the World Bank, Washington, D.C., 2 & 3 April, 1992.
König, Dieter (1995): Sustainable Development: Linking Global Environmental Change to Technology Co-operation, in: Dwivedi, O. S. and Dhirendra K. Vajpeyi (ed.): Environmental Policies in the Third World: a Comparative Analysis, Westport, p. 1-21.
Kohl, Helmut (1992): Rede auf der UN-Konferenz Umwelt und Entwicklung in Rio de Janeiro, 12 Juni 1992.
Kohler-Koch, Beate (1993): Die Welt regieren ohne Weltregierung, in: Böhret, Carl & Göttrik Wewer (eds.): Regieren im 21. Jahrhundert – zwischen Globalisierung und Regionalisierung, Pladen, p. 109-141.
Kokusai Kaihatsu Jaanaru (International Development Journal) (1996): February, No. 471.
Kokusai Kaihatsu Jaanaru (1995): February, No. 458.

Korhonen, Pekka (1994): The Theory of the Flying Geese Pattern of Development and its Interpretations, in: Journal of Peace Research, Vol. 31, p. 93-108.
Kouzmin, Alexander & Andrew Hayne (1999): Globalization: Rhetorical Trends and Unpalatable Realities For Public Sectors, in: Kouzmin, Alexander & Andrew Hayne: Essays in Economic Globalization, Transnational Policies and Vulnerability, Amsterdam, p. 1-12.
Kritz, Jürgen, Helmut E. Lück und Horst Heidbrink (1990): Wissenschafts- und Erkenntnistheorie, Opladen, 2. Edition.
Kuehr, Ruediger (2007): Environmental technologies – from misleading interpretations to an operational categorisation & definition, in: Journal of Cleaner Production, Vol 15 (2007), p. 1316-1320.
Kühr, Rüdiger (1996): Coping with Tokyo's Mountain of Waste. Recycling – A Solution?, in: Japan Environment Monitor, August-September, p. 4-6.
Kühr, Rüdiger (1997): Japans umweltpolitische Entwicklungshilfe. Eine Analyse am Beispiel des Transfers von Umwelttechnologie, in: Pohl, Manfred (ed.) Japan 1996/97, Politik und Wirtschaft, Hamburg, p. 199-222.
Kühr, Rüdiger and György Széll (eds.) (1997): „Zero Emission" – eine Chance in Europa? Dokumentation des 1. Europäischen Zero Emission Kongresses, 23./24. Oktober 1997, Osnabrück.

Lehner, Franz & Friedrich Schmidt-Bleek (1999): Die Wachstums-maschine. Der ökonomische Charme der Ökologie, München.
Less, Cristina & Steven McMillan (2005): Achieving the Successful Transfer of Environmentally Sound Technologies: Trade-Related Aspects, OECD Trade and Environment Working Paper, 2005-2.
Lindblom, Charles F. (1980): The Policy-Making Process, New Jersey.
Löbler, Frank (1990): Stand und Perspektiven der Policy-Forschung in der deutschen Politikwissenschaft. Ein empirisches Profil des Forschungsgegenstandes, Siegen.
Löbler, Frank (1988): Stand und Perspektiven der Politikfeldanalyse in der deutschen Politikwissenschaft, Bochum.
Lovelock, James (1993): The Evolving Gaia Theory, paper presented at the United Nations University on 25 September 1992, Tokyo.
Luhmann, Niklas (1988a): Die Wirtschaft der Gesellschaft, Frankfurt/Main.
Luhmann, Niklas (1988b): Ökologische Kommunikation. Kann die moderne Gesellschaft sich auf ökologische Gefährdungen einstellen? Opladen, 2^d Edition.
Lund University (ed.) (1996): Donor-funded Cleaner Production Programmes in Developing Countries, by Bardouille, Pepukaye.

Maddock, Rowland T. (1994): Japan and Global Environmental Leadership, in: Journal of Northeast Asian Studies, Vol. XIII, No. 4, p. 37-48.
Mallett, Ed (2010): The World Environmental Market, in: EcDevJournal.com (http://www.ecdevjournal.com/index.php?option=com_content&task=view&id=141&Itemid=29 – 14 July 2010).
Mankiw, Gregory (2007): Principles of Economics, Cincinnati.
Mann, Michael (1986): The Source of Social Power, Cambridge.
Maruko, Maya (1995): Nations look to Japanese firms to help economic expansion, in: The Japan Times, 04 February 1995, p. 8.
Matsuoka, Shunji (1996): Japan's ODA and Environment Co-operation in Southeast Asia, in: Journal of International Development and Co-operation, Vol. 2, No. 1, p. 35-57.
Maull, Hanns W. (1992): Japan's Global Environmental Policies, in: The international politics of the environment, Oxford, p. 354-372.
Maull, Hanns W. (1991): Japans internationale Umweltpolitik, in: Europa-Archiv, p. 503-512.
May, Bernhard (1989a): Die neue Entwicklungspolitik Japans, in: Asien, Nr. 30, p. 40-59.
May, Bernhard (1989b): Japans neue Entwicklungspolitik. Entwicklungshilfe und japanische Außenpolitik, München.
Maybury, Robert (ed.) (1986): Violent Forces of Nature, Mt. Airy.
Mayntz, Renate et al. (1978): Vollzugsprobleme der Umweltpolitik, Stuttgart.
Mayntz, Renate (1979): Regulative Politik in der Krise?, in: Matthes, Joachim (ed.): Sozialer Wandel in Westeuropa. Verhandlungen des 19. Soziologentages in Berlin, Frankfurt, p. 55-81.
Mayntz, Renate (1982): Problemverarbeitung durch das politisch-administrative System: Zum Stand der Forschung, in: Hesse, Joachim J. (ed.): Politikwissenschaft und Verwaltungswissenschaft, Opladen, p. 74-89.
Mayntz, Renate (1987): Politische Steuerung und gesellschaftliche Steuerungsproblem – Anmerkungen zu einem theoretischen Paradigma, in: Ellwein Thomas et al. (eds.): Jahrbuch zur Staats- und Verwaltungswissenschaft, Vol. 1, Baden Baden, p. 89-110.
Mayntz, Renate (1988): Funktionelle Teilsystems in der Theorie sozialer Differenzierung, in: Mayntz, Renate et al. (eds.): Differenzierung und Verselbstständigung. Zur Entwicklung gesellschaftlicher Teilsystems, Frankfurt, p. 11-44.
McKean, Margaret (1981): Environmental Protest and Citizen Politics in Japan, Berkeley.
McKibbin, Warwick (1998): Modelling the crisis in Asia, in: ASEAN Economic Bulletin, December.

Meadows, Dennis et al. (1972): The Limits to Growth, New York.
Mendel, Wolf (1978): Issues in Japan's China Policy, London
Menzel, Ulrich (1998) Das Ende der einen Welt und die Unzulänglichkeit der kleinen Theorien, in: Entwicklung und Zusammenarbeit 39.2, p. 45-48.
Menzel, Ulrich (1991) Das Ende der „Dritten Welt" und das Scheitern der großen Theorie. Zur Soziologie einer Disziplin in auch selbstkritischer Absicht, in: Politische Vierteljahresschrift 32, p. 4-33.
Messner, Dirk (1998): Die Transformation von Staat und Politik im Globalisierungsprozess, in: Messer, Dirk (ed.): Die Zukunft des Staates und der Politik. Möglichkeiten und Grenzen politischer Steuerung in der Weltgesellschaft, Bonn, p. 14-43.
Meves, Hans (1993): Japanese Environmental Policy – Alternating Stimulus and Abstinence, in: Deutsches Institut für Japanstudien der Philipp-Franz-von-Siebold-Stiftung: Japanstudien, Jahrbuch, Band 4, München, p. 155-182.
Meyers, Reinhard (1991): Grundbegriffe, Strukturen und theoretische Perspektiven der Internationalen Beziehungen, in: Bundeszentrale für politische Bildung: Grundwissen Politik, Bonn, p. 220-316.
Ministry of Foreign Affairs (2010): Japan's Official Development Assistance, White Paper 2009, Japan's International Cooperation, Tokyo.
Ministry of Foreign Affairs (2002): Koizumi Initiative (Concrete Actions of Japanese Government to be taken for Sustainable Development - Towards Global Sharing), 21 August 2002.
Ministry of Foreign Affairs (2000): Diplomatic Bluebook, Tokyo.
Ministry of Foreign Affairs (2000): Japan's Official Development Assistance (ODA) Annual Report 1999, Tokyo.
Ministry of Foreign Affairs (1996): Japan's Official Development Assistance (ODA) Annual Report 1995, Tokyo.
Ministry of Foreign Affairs (1995a): Japan's Official Development Assistance (ODA) Annual Report 1994, Tokyo.
Ministry of Foreign Affairs (1995b): Japan's Official Development Assistance Summary 1995, Tokyo.
Ministry of Foreign Affairs (1990): Waga kuni no Seifu Kaihatsu Enjo, Vol. 1, Tokyo.
Ministry of International Trade and Industry (MITI) (1995a): Japan's Economic Co-operation 1994, The Expansion of Dynamism in Asia, Tokyo.
Ministry of International Trade and Industry (1995b): Green Aid Plan (project for 1995) in Indonesia, Tokyo 1995.
Ministry of International Trade and Industry (1992): Fourteen Proposals for a New Earth, Policy Triad for the Environment, Economy and Energy, Executive Summary, Tokyo, 25 November 1992.

Ministry of International Trade and Industry (1990): Waga Kuni no Seifu Kaihatsu Enjo (Japan's Official Development Assistance), Vol. 1, Tokyo.
Mishan, Edward J. (1969): Growth: The price we pay, London.
Montes, Manuel F. (1998): The Currency Crisis in Southeast Asia. Singapore, Institute of Southeast Asian Studies.
Morse, Edward L. (1970): The Transformation of Foreign Policies: Modernization, Interdependence, and Externalization, in: Little, Richard & Michael Smiths (eds.): Perspectives on World Politics, 2nd ed. (this article is a reprint and was originally published in World Politics 22/3 in 1970, p. 371-392).
Moore, Curtis and Alan Miller (1994a): Green Gold, Japan, Germany, the United States, and the Race for Environmental Technology, Boston.
Moore, Curtis and Alan Miller (1994b): Strenghts And Limitations of Governmental Support For Environmental Technology in Japan, in: Industrial & Environmental Crisis Quarterly, Vol. 8, No. 2, p. 155-169.
Moriguchi, Kenzo (1995): Japanese aid said to benefit Indonesia companies, in: The Japan Times, 10. Juni 1995, p. 5.
Münckler, Herfried et al. (2002): Der demokratische Nationalstaat in Zeiten der Globalisierung. Politische Leitideen für das 21ste Jahrhundert, Festschrift zum 80. Geburtstag von Iring Fetscher, Berlin.
Murdo, Pat (1994): Japan's Environmental Foreign Aid: What Kind of Edge? in: Japan Economic Institute (ed.): JEI-Report, No. 31 A, Washington, p. 1-20.
Mulgan, Geoff (1995): Myth of Withering Government, in: The Independent, 15. May.
Mutius, Albert von (ed.) (1992): Lorenz von Stein. 1890–1990, Akademischer Festakt zum 100. Todestag. Verlag von Decker, Heidelberg.

Nagasaki, Fukuzo. (1995). Research on whales. Tokyo.
Nashima, Mitsuko (1995): Nation urged to end inaction. Japan urged to champion rights in Asia, in: The Japan Times, p. 6.
Naßmacher, Hiltrud (1991): Vergleichende Politikforschung. Eine Einführung in Probleme und Methoden, Opladen.
NationMaster.com (2010): GDP (1978) by country (http://www.nationmaster.com/graph/eco_gdp-economy-gdp&date=1978 – 17 July 2010).
Natori, Makoto (1993): Japan's Pollution Control Technologies and Their Role in the World, in: Japan Review of International Affairs, Winter, p.50-67.
Naturschutzbund Deutschland e.V. (NABU) (1999): 100 Jahre NABU – Historischer Abriss, Bonn.

New Energy Development Organisation (NEDO) (1993): The Innovation of New Technology, Tokyo.
Nohlen, Dieter (1995): Wörterbuch Staat und Politik; Bundeszentrale für politische Bildung, Bonn.
Nordisk Innovations Center (2010): Clean, Clever and Competitive, Nordic Environmental Technologies, Oslo.
Nose, Yasunobu (1994): Polluted city transfers "green" know-how. Foreign trainees learn how to protect, improve environment, in: The Nikkei Weekly, 06. Juni 1994, p. 21.
Nuscheler, Franz (2001): Warum brauchen wir Entwicklungstheorien, in: Thiel, Reinhold E. (ed.): Neue Ansätze zur Entwicklungstheorie. Deutsche Stiftung für Entwicklung (DSE), p. 389-399.
Nuscheler, Franz (1995): Lern- und Arbeitsbuch Entwicklungspolitik, Bonn, 4th Edition.
Nuscheler, Franz (1994): Japan als „aid leader": Neue Entwicklungen in der japanischen Entwicklungspolitik, in: Pohl, Manfred (ed.): Japan 1993/94. Politik und Wirtschaft. Hamburg, p. 163-180.
Nuscheler, Franz (1991): Lern- und Arbeitsbuch Entwicklungspolitik, Bonn, 3^{d} Edition.
Nuscheler, Franz (1990): Japans Entwicklungspolitik. Quantitative Superlative und qualitative Defizite, Hamburg.

O'Connor, David (1996): Grow Now/Clean Later, or Pursuit of Sustainable Development? Paris.
O'Connor, David (1994): Managing the Environment with Rapid Industrialisation. Lessons from the East Asian Experience, Paris.
OECD (2005): Environmental Data Compendium 2004, Paris
OECD (2002): Development Co-operation Review Germany: 2001.
OECD (1999a): Environmental Data Compendium 1999, Paris.
OECD (1999b): Development Co-operation Review Germany: 1998 No. 34, Paris.
OECD (1999c): Development Co-operation Review Japan: 1999 No. 34, Paris.
OECD (1997): Environmental Data Compendium 1997, Paris.
OECD (1996): Development Co-operation Review Series. Japan 1996, No. 13, Paris.
OECD (1995): Promoting Cleaner Production in Developing Countries, The Role of Development Co-operation, Paris.
OECD (1992): The OECD environment industry: Situation, prospects and government policies, Paris.
OECD (1994): Environmental Performance Reviews: Japan, Paris.

OECD (1988): The Newly Industrialising countries. Challenge and Opportunity for OECD Industries, Paris.
OECD (1985): Recommendation of the Council on Environmental Assessment of Development Assistance Projects and Programmes, C (85)104, Paris.
OECF (1995): Annual Report 1995, Tokyo.
OECF (1994): Annual Report 1994, Tokyo.
OECF (1990): Annual Report 1990, Tokyo.
Okita, Saburo (1981): The Developing Economies and Japan: Lessons in Growth, Tokyo.
Ott, Konrad & Ralf Döring (2004): Theorie und Praxis starker Nachhaltigkeit, Marburg.

Park, Jacob (1995): Japanese Policy on Climate Change, in: Social Science Japan, August, p. 32.
Pasuk, Phongpaichit & Chris Baker (2000): Thailand's Crisis. Chiang Mai.
Paucke, Horst & Günter Streibel (1990): Ökonomie contra Ökologie? Ein Problem unserer Zeit, Berlin.
Paucke, Horst (1996): Ökologisches Erbe und ökologische Hinterlassenschaft, Marburg.
Paucke, Horst (1994): Chancen für Umweltpolitik und Umweltforschung. Zur Situation in der ehemaligen DDR, Marburg.
Pauli, Gunter A. (1987): Crusader for the Future: A Portrait of Aurelio Peccei, Founder of the Club of Rome, Oxford.
Peters, B. Guy (1998): Comparative Politics. Theory and Methods, Houndmills.
Pleticha, Heinrich (Ed.) (1987): Deutsche Geschichte. Von der „Restauration bis zur Reichsgründung" 1815-1871, Gütersloh.
Pollack, Andrew (1992): Ecological Savior Abroad, Japan Lags at Home, in: International Herald Tribune, 01 August 1992, p. 11.
Popp, David (2009): Policies for the Development and Transfer of Eco-Innovations: Lessons from the Literature, OECD Environment Working Papers, No. 10.
Popper, Karl R. (1976): Logik der Forschung, Tübingen.
Potter, David (1994): Assessing Japan's environmental aid policy, in: Pacific Affairs, Vol. 67, p. 200-215.
Prittwitz, Volker von (1994): Politikanalyse, Opladen.
Prittwitz, Volker von (1993): Katastrophenparadox und Handlungskapazität. Theoretische Orientierungen der Politikanalyse, in: Héritier, Adrienne (Hrsg.): Policy-Analyse. Kritik und Neuorientierung, Opladen.
Prittwitz, Volker von (1992): Methodische und theoretische Grundpositionen der Umweltpolitik, in: Dreyhaupt, Franz Joseph, Franz-Joseph Peine und

Gerhard W. Wittkämper (eds.): Umwelt Handwörterbuch, Berlin, p. 163-168.
Prittwitz, Volker von (1990): Das Katastrophenparadox. Elemente einer Theorie der Umweltpolitik, Opladen.
Putnam, Robert D. & Nicholas Bayne (1985): Weltwirtschaftsgipfel im Wandel, Bonn.

Ramage, Douglas E.(1995): Politics in Indonesia : democracy, Islam and the ideology of tolerance, London.
Rahmstorf, Stefan & Hans-Joachim Schellnhuber (2007): Der Klimawandel, München.
Rasiah, Rajah (2001): Pre-crisis economic weaknesses and vulnerabilities, in: Jomo KS, ed. Malaysian Eclipse: Economic Crisis and Recovery. London, p. 47–66.
Rat von Sachverständigen für Umweltfragen (1987): Umweltgutachten 1987, Bonn, BT 11/1568.
Rau, Theo (2009): Entwicklungspolitik, Braunschweig.
Recktenwald, Horst C. (1978): Unwirtschaftlichkeit im Staatssektor – Elemente einer Theorie des ökonomischen Staats"versagens", in Hamburger Jahrbuch für Wirtschafts- und Gesellschaftspolitik, Tübingen, p. 155-165.
Reed, Steven (1986): Japanese Prefectures and Policymaking, Pittsburgh.
Regional Institute of Environmental Technology (RIET) (Ed.) (1996): The Asian Environmental Market. An Overview of Business Opportunities, Singapore.
Reischauer, Edwin O. (1952): Japan, Berlin.
Reid, David (1995): Sustainable Development – An Introductory Guide, London.
Richter, Emanuel (1992): Der Zerfall der Welteinheit. Vernunft und Globalisierung der Moderne, Frankfurt.
Risse-Kappen, Thomas (1995): Bringing Transnational Relations Back In: Introduction, in: Risse-Kappen, Thomas (ed.): Bringing Transnational Relations Back In: Non-State Actors, Domestic Structures and International Institutions, Cambridge, p. 3-33.
Rittberger, Volker und Klaus Dieter Wolff (1985): Policy-Forschung und Internationale Beziehungen, in: Hartwich, Hans-Hermann (ed.): Policy-Forschung in der Bundesrepublik Deutschland. Ihr Selbstverständnis und ihr Verhältnis zu den Grundfragen der Politikwissenschaft, Opladen, p. 204-211.

Robert, Rüdiger (1988): Politikwissenschaft und Politikbegriffe, in: Bellers, Jürgen und Rüdiger Robert (eds.): Politikwissenschaft 1 - Grundkurs, Münster, p. 1-29.

Robèrt, Karl-Henrik et al. (2004) Strategic Leadership towards Sustainability, Karlskrona.

Robertson, Roland (1992): Globalization: Social Theory and Global Culture, London.

Robertson, Roland (1990): Mapping the Global Conditions, in: Theory, Culture and Society, Vol. 7, No.: 2, p. 63-99.

Robinson, John & John Tinker (1997): Reconciling Ecological, Economic and Social Imperatives: A New Conceptual Framework, in: Schrecker, Ted (ed.): Surviving Globalism, The Social and Environmental Challenges, London, p. 71-94.

Rock, Michael (1994): Transitional democracies and the shift to export-led industrialisation: lessons from Thailand, in: Studies in Comparative International Development, Spring.

Rohde, Miriam (2003): Japans Entwicklungszusammenarbeit. Auf dem Weg zu Good Governance, Mitteilungen des Instituts für Asienkunde, Hamburg.

Rohde, Miriam (1996): Japans staatliche Entwicklungshilfe (ODA): Ansätze zu einer neuen Entwicklungspolitik? in: Japan (Wirtschaft, Politik, Gesellschaft), p. 416-427.

Rohde, Miriam (1995a): Japans staatliche Entwicklungshilfe (ODA). Struktur, Entwicklung und Perspektiven, in: Japan (Wirtschaft, Politik, Gesellschaft), August, p. 390-400.

Rohde, Miriam (1995b): Japans staatliche Entwicklungshilfe (ODA). Verflechtungen zwischen staatlichem und privatem Sektor, in: Japan (Wirtschaft, Politik, Gesellschaft), p. 601-612.

Ronge, Volker (ed.) (1980): Am Staat vorbei?, Frankfurt/M.

Ropohl, Günter (1991): Technologische Aufklärung: Beiträge zur Technikphilosophie, Frankfurt/Main.

Rorty, Richard (1994): Towards a Liberal Utopia, in: Times Literary Supplement, 24 June.

Rosenau, James N. (1990): Turbulence in world Politics. A theory of change and continuity, Princeton.

Rosenau, James N. (1997): Along the Domestic-Foreign Frontier: Exploring Governance in a Turbulent World, Cambridge.

Rosenau, James N. & Erst-Otto Czempiel (eds.) (1992): Governance Without Government: Order and Change in World Politics, New York.

Ross, George et al. (2002): Walled-in Summits vs. Parliaments in the Streets : New Paths Toward Global Democracy? (http://www.planetagora.org/english/article.html – 30 July 2003).

Rousseau, Jean-Jacques (1974): The Social Contract or Principles of Political Right, plus the Dedication from the „Second Discourse" and on Political Economy, New York.

Roy, Ajit (1995): Civil Society and nation State. In Context of Globalisation, in: Economic and Political Weekly, August 5-12, p. 2005-2011.

Rüschemeyer, Dietrich (1974): Reflections on Structural Differentiation, in: Zeitschrift für Soziologie 3, p. 279-294.

Ruschkowski, Eick (2002): Lokale Agenda 21 in Deutschland – eine Bilanz, in: Aus Politik und Zeitgeschichte, B 21/-32, p. 17-24.

Rüschemeyer, Dietrich (1977): Structural Differentiation, Efficiency, and Power, in: American Journal of Sociology 83, p. 1-25.

Sargoff, Mark (1988): The Economy of the Earth. Philosophy, Law and the Environment, Cambridge.

Sarkar, Saral (1993): Green-Alternative Politics in West Germany, The New Social Movements, Vol. I, Tokyo.

Scharpf, Fritz W. (1998): Demokratie in der transnationalen Politik, in: Beck, Ulrich (Ed.): Politik der Globalisierung, Frankfurt, p. 228-253

Scharpf, Fritz W. (1989): Politische Steuerung und politische Institutionen, in: Hartwich, Hans-Hermann (ed): Macht und Ohnmacht politischer Institutionen, Opladen, p. 17-29.

Scheleman, Ferd *(1998).* Evaluation of the International Environmental Technology Centre of the United Nations Environment Programme, Nairobi.

Schimank, Uwe (1992): The Worsening of Research Conditions at German Universities: Individual Coping Makes the Best of Corporate Coping's Failure. Discussion Paper presented during the conference "Coping with Trouble: How Scientists and Research Institutes React to Political Disturbances of their Research Condition", Köln.

Schmitt, Uwe (1991): Hilf dir selbst, dann hilft dir Japan, in: Frankfurter Allgemeine Zeitung, 28 September.

Schrecker, Ted (1997): Introduction: Sustainability, Globalization and Moral Imagination, in: Schrecker, Ted (ed.): Surviving Globalism, The Social and Environmental Challenges, New York, p. 1-12.

Schreiber, Helmut (1986): Grenzüberschreitende Umweltprobleme zwischen der Volksrepublik Polen und der DDR, in: Haendcke-Hoppe, Maria & Konrad Merkel (eds.): Umweltschutz in beiden Teilen Deutschlands, Berlin, p. 131-143.

Schreurs, Miranda A. (2009): Germany's Environmental Transformation: From Pollution Haven to Environmental Leader, in: AICGS Transatlantic Perspectives, December 2009, p. 1-4.
Schreurs, Miranda A. (1997): A political system's capacity for global environmental leadership: A case study of Japan, in: Mez, Lutz & Helmut Weidner (eds.): Umweltpolitik und Staatsversagen. Perspektiven und Grenzen der Umweltpolitikanalyse, Berlin, p. 323-331.
Schreurs, Miranda A. (1994): Nihon ni okeru kankyō seisaku no kettei katei (Policy Laggard or Policy Leader. Global Environmental Policy Making Under the Liberal Democratic Party), in: Journal of Pacific Asia, Vol. 2, p. 3-38.
Schröder, Gerhard (2002): Statement by Mr. Gerhard Schröder Chancellor of the Federal Republic of Germany at the World Summit on Sustainable Development, Johannesburg, South Africa, 2 September 2002.
Schubert, Klaus (1991): Politikfeldanalyse, Opladen.
Schubert, Bernd et al. (1984). Die Nachhaltigkeit der Wirkungen von Agrarprojekten – Eine Querschnittsanalyse von Projekten der deutschen Technischen Zusammenarbeit, Forschungsberichte des BMZ, Bonn.
Schumacher, Ernst. F. (1973): Small is Beautiful: Economics as if People Mattered, New York.
Shaw, Martin (1997): The state of globalization: towards a theory of state transformation, in: Review of International Political Economy, Autumn 1997, p. 497-513.
Shen, Thomas (1995): Industrial Pollution Prevention, Berlin.
Simonis, Udo Ernst (2002b): Wer rettet die globale Ökologie? Plädoyer für eine Weltorganisation für Umwelt und Entwicklung, in: Münckler, Herfried et al.(eds.): Der demokratische Nationalstaat in Zeiten der Globalisierung. Politische Leitideen für das 21ste Jahrhundert, Festschrift zum 80. Geburtstag von Iring Fetscher, Berlin, p. 73-87.
Simonis, Udo Ernst (2002a): Globale Umweltprobleme und die neue Weltpolitik, in: Kaiser, Karl & Hans-Peter Schwarz (eds.): Weltpolitik im neuen Jahrhundert, p.137-149.
Simonis, Udo Ernst et al. (1999): Weltumweltpolitik. Grundriss und Bausteine eines neuen Politikfeldes, Berlin.
Simonis, Udo Ernst (1996): Globale Umweltpolitik: Ansätze und Perspektiven, Mannheim.
Simonis, Udo Ernst (ed.) (1984): Mehr Technik – weniger Arbeit? Plädoyer für sozial- und umweltverträgliche Technologien, Karlsruhe.
Skare, Mari (1994). Whaling: a sustainable use of natural resources or a violation of animal rights? In: Environment, vol. 36, p. 12-26.

Singer, J. David (1969): The Global System and its Subsystems: A development view, in: James N. Rosenau (ed.): Linkage Politics: Essays on the Convergence of National and International Systems, New York, p. 123-134.
Sklar, Holly (1980): Trilaterialism. The Trilateral Commission and Elite Planning for World Management, Boston.
Smith, Kirk R. (1993): The basic greenhouse gas indices, in: Hayes, Peter & Kirk Smith (eds.): The global greenhouse regime. Who pays? Tokyo, p. 20-50.
Smith, Fraser (1997): A synthetic Framework and a Heuristic for Integrating Multiple Perspectives on Sustainability, in: Smith, Fraser (ed.): Environmental Sustainability, Practical Global Implications, Boca Raton, p. 1- 24.
Solar Promotion Association (SPA) (2003): Bevölkerung wünscht Vorreiterrolle Deutschlands, Emnid-Umfrage zur Erneuerbaren Energie im Kontext von Klimaveränderung und Unwetterkatastrophe (http://www.sfv.de/lokal/mails/wvf/ vorreite.htm – 27 January 2004).
Spöhring, Walter (1989): Qualitative Sozialforschung, Stuttgart.
Statistisches Bundesamt (ed.): Datenreport 1994. Zahlen und Fakten über die Bundesrepublik Deutschland, Bonn (Schriftenreihe der Bundeszentrale für politische Bildung, Band 325)
Stern, David I. (2004): The rise and fall of the environmental Kuznets curve, in: World Development, 32(8), p. 1419-1439.
Stern, David I. & R. K. Kaufmann (1996): Estimates of global anthropogenic methane emissions 1860-1993, in: Chemosphere No. 33, p. 159-176.
Strong, Maurice F. (1992): Themen für den Umweltgipfel 1992 – die globale Herausforderung, in: Europa-Archiv, Folge 9, p. 231-237.
Strübel, Michael (1992): Internationale Umweltpolitik, in: Dreyhaupt, Franz Joseph, Franz-Joseph Peine und Gerhard W. Wittkämper (eds.): Umwelt Handwörterbuch, Berlin, p. 222-225.
Sullivan, Michael J. (1996): Comparing state polities: a framework for analyzing 100 governments, Westport.
Széll, György & Ute Széll (eds.) (2009) Quality of Life & Working Life in Comparison. Frankfurt/M.
Széll, György & Ken'ichi Tominaga (eds.) (2004): Environmental Challenges for Japan and Germany – Interdisciplinary and Intercultural Perspectives, Frankfurt/M.

Taniguchi, Masaki (2008): The State of Political Science in Japan, Tokyo.
Thiel, Reinold (ed.) (2001): Neue Ansätze zur Entwicklungstheorie, Bonn: DSE/IZEP, 2nd Edition, p. 9-34.

Tokyo Conference on Global Environmental Action (1994): Issue Papers, Tokyo.
Torgerson, Douglas (1995): The uncertain quest of sustainability: public discourse and the politics of environmentalism, in: Fischer, Frank & Michael Black (eds.): Greening Environmental Policy: The Politics of a Sustainable Future, London, p. 3-20.
Tsuru, Shigeto (1999): The Political Economy of the Environment, The Case of Japan, London.
Tsuru, Shigeto und Helmut Weidner (eds.) (1989): Environmental Policy in Japan, Berlin.

UBA (2001): www.umweltbundesamt.org/dzu/Y00269.html – 15 May 2010
Ueta, Kazuhiro (1998): Chikyu kankyōhozen to nakyo gijutsu iten, in: Takahashi, Hiroshi & Kazuhiko Takeuchi (eds.): Chikyu kankyō kagaku. Chikyu shisutemu wo sasaeru 21seiki gata kagaku gifutsu, Tokyo.
Uexküll, Gudrun von (1964): Jakob von Uexküll. Seine Welt und seine Umwelt; eine Biographie, Hamburg.
Ui, Jun (1992): Minamata Disease, in: Ui, Jun (ed.): Industrial Pollution in Japan, Tokyo, p. 103-132.
Ui, Jun (1996): Nihon no mizu wa yomigaeru ka?
Ui, Jun (1993): Kōgai genron, Tokyo, 4th Edition.
Uitto, Juha I. (1994).: Environment, Technology And The Japanese Experience Lessons For Developing Countries From International Collaborative Research, in: Industrial & Environmental Crisis Quarterly, Vol. 8., No. 2.
UNEP (2005): Bali Strategic Plan for Technology Support and Capacity-building, Nairobi.
UNEP (1998): UNEP Division for Technology, Industry and Economics Activity Report 1998, Paris.
United Nations General Assembly (2004): Resolution 57/253, New York.
United Nations General Assembly (1997): Nineteenth Special Session, 1st plenary meeting, Monday, New York, 23 June 1997.
United Nations (UN) (2004): The Johannesburg Summit Test: What Will Change? (http://www.johannesburgsummit.org/html/whats_new/feature_story41.html – 12 November 2004).
United Nations (UN) (1994): Human Development, Institutions and Technology Branch. Division for Sustainable Development. Department for Policy Co-ordination and Sustainable Development: Agenda 2 1: Technology Related Issues, New York.
United Nations (1992): Environmentally Sound Technology for Sustainable Development, New York.

UNCED (1992): Agenda 21 (advanced version as adopted by the Plenary in Rio de Janeiro, 14 June 1992), Geneva.
UNDP (1997): Governance for sustainable human development. A UNDP policy document. New York.
UNDP (1994): Human Development Report, New York.
UNDP (1998): Human Development Report, New York.
UNDP (2004): Human Development Report, New York.
UNFCCC (2010): Kyoto Protocol, Bonn.
Universität Kaiserslautern (ed.) (1997): Volkswirtschaftliche Diskussionsbeiträge. The Emerging Market of Environmental Technology in Asia by Hauff, Michael and Martin Wilderer, Kaiserslautern.
Unti, Bernard (2010): IWC 2010: Final Dispatch, Deal dead, moratorium intact, future uncertain, in: Human Society International, 25 June 2010.
U. S. Congress, Office of Technology Assessment (1993): Development Assistance, Export Promotion, and Environmental Technology, Washington D.C.
U. S. International Trade Commission East Asia (1993): Regional Economic Integration and Implications for the United States, Washington D.C.

Vollmer, Klaus (ed.) (2006): Environmental policies and ecological issues in Japan and Eastern Asia, München.
Voppe, Götz (1990): Die Industrialisierung der Erde, Stuttgart.

Wackerbauer, Johan (1995): Struktur und Wettbewerbssituation der Anbieter von Umwelttechnik und unweltfreundlicher Technik, in: Ifo Schnelldienst, Ifo Institute for Economic Research at the University of Munich, Vol. 48 (21), p. 7-14.
Ward, Lester F. (1907): Reine Soziologie. Eine Abhandlung über den Ursprung und die spontane Entwicklung der Gesellschaft, Innsbruck.
Waters, Malcom (1995): Globalization, London.
Waugh, David (2000a): World development, in: Geography, An Integrated Approach.
Waugh, David (2000b): Manufacturing industries, in: Geography, An Integrated Approach.
Weber, Max (1991): Wirtschaftsgeschichte: Abriss der universalen Sozial- und Wirtschaftsgeschichte, Berlin.
Weber, Max (2000): Die protestantische Ethik und der Geist des Kapitalismus, Hamburg.
Weber, Max (1990): Briefe 1906-1908, Tübingen (edited by Lepsius M. Rainer & Wolfgang J. Mommsen).

Weidner, Helmut (1996): Basiselemente einer erfolgreichen Umweltpolitik. Eine Analyse und Evaluation der Instrumente der japanischen Umweltpolitik, Berlin.

Weidner, Helmut (1988): Bausteine einer präventiven Umweltpolitik, Anregungen aus Japan, in: Simonis, Udo E. (ed.): Präventive Umweltpolitik, Frankfurt/Main, p. 143-166.

Weidner, Helmut; Eckard Rehbinder und Rolf-Ulrich Sprenger (1990): Die Umweltpolitik in Japan: Ein Modell für die EG?, in: ifo-Schnelldienst, 1617/1990, p. 35.

Weismantel, Wolfgang (1987): Anfänge der Industrialisierung und der sozialen Frage, in: Pleticha, Heinrich (ed.): Von der „Restauration" bis zur Reichsgründung 1815-1871, Gütersloh, p.172-199.

Weiß, Jens (2000): Umweltpolitik als Akteurshandeln, Marburg

Weizsäcker, Ernst Ulrich von (1997): Erdpolitik, Darmstadt, 5th Edition.

Weizsäcker, Ernst Ulrich von et al. (1996): Faktor Vier. Doppelter Wohlstand – halbierter Naturverbrauch, München.

Wey, Klaus Georg (1982): Umweltpolitik in Deutschland: kurze Geschichte des Umweltschutzes in Deutschland seit 1900, Opladen.

White, Gilbert F. (1996): Emerging Issues in Global Environmental Policy, in: AMBIO, a journal of the human environment, Vol. 25, No. 1, February, p. 58-60.

Wicke, Lutz (1982): Umweltökonomie. Eine praxisorientiere Einführung, München.

Wild, A. (1994): Vision of 2022: Era of the Global Works Council and Individualism, in: Personal Management, Vol. 26, No. 13, p. 39-49.

Willke, Helmut (1996):Ironie des Staates: Grundlinien einer Staatstheorie polyzentrischer Gesellschaft, Frankfurt/Main.

Willke, Helmut (1989): Systemtheorie entwickelter Gesellschaften. Dynamik und Riskanz moderner gesellschaftlicher Selbstorganisation. Weinheim.

Windhoff-Héritier, Adrienne (1987): Policy-Analyse. Eine Einführung, Frankfurt/Main.

Whitehead, J. Rennie (2004): A Brief History of the Club of Rome – A Summary and Personal Reminiscences (http://www3.sympatico.ca/drrennie/corhis.html#ClubofRome – 6 December 2004).

Wirtschaftministerium des Landes Baden-Württemberg (Ed.) (1992): Umwelttechnologien in Ostasien. Länderdarstellungen, Strategien, Finanzierungs- u. Förderprogramme. Analyse des Kooperationspotentials für baden-württembergische Unternehmen im Bereich des Umweltschutzes in Asien (ausgewählte Länder) 1993-2005, Stuttgart.

Wissenschaftlichen Beirats beim Bundesministerium der Finanzen (2010): Klimapolitik zwischen Emissionsvermeidung und Anpassung, Berlin, Januar 2010.
Wittkämper, Gerhard (1992): Umweltschutz. Einführung in Umweltpolitik und Umweltverwaltung unter Berücksichtigung des Umweltrechts, Regensburg.
World Bank Metropolitan Environmental Improvement Program (MEIP) (1995): Japan's Experience in Urban Environmental Management, Washington D.C.
World Bank (1997): Environment Matters, Washington.
World Bank (1994): Indonesia: Environment and Development. A World Bank Country Study, Washington D.C.
World Bank (1993): The East Asian Miracle: Economic Growth and Public Policy, New York.
World Bank (1992): World Development Report 1992, Washington.
World Bank (2010): World Development Indicators, Washington.
World Commission on Environment and Development (WCED) (1987): Our Common Future, Oxford.

Yamada, Michiaki (1995a): Yen loans to China remain focus of debate, in: The Daily Yomiuri, 05 January 1995, p. 12.
Yamada, Isao (1995b): Fulfilling Japan's responsibility as the world's foremost aid donor nation. Interview with Kimio Fujita, President, Japan International Co-operation Agency, in: Japan 21st, Vol. 40, p. 12-13.
Yamamoto, Wataru (1994): Japanese Official Development Assistance and Industrial Environmental Management in Asia. Workshop on Trade and Environment in Asia-Pacific: Prospects for Regional Co-operation. Japan Development Institute (JDI), Tokyo.
Yook, Soon-Hyung et al. (2001): Modelling the Internet's large-scale topology, in: PNAS, 15 October 2002, Vol. 99, No. 21, p. 13382-13386.
Young, Oran (ed.) (1997): Global Governance: Drawing Insights from the Environmental Experience, Cambridge.
Young, Oran (1990): Global environmental change and international governance, in: Journal of International Studies, 19:3, p. 337-346.

Zacher, Mark W. (1992): The Decaying Pillars of the Westphalian Temple: Implications for International Order and Governance, in: Rosenau, James N. & Ernst-Otto Czempiel (eds.): Governance without Government: Order and Change in World Politics, Cambridge, p. 58-73.

Zürn, Michael (1998a): Schwarz-Rot-Grün-Braun: Reaktionsweisen auf Denationalisierung, in: Beck, Ulrich (ed.): Politik der Globalisierung, Frankfurt, p. 297-330.

17.2 Internet

http://www.whaling.jp/english/history.html – 28 March 2010.
http://www.iwcoffice.org – 01 July 2010.
http://www.toyota-media.de – 01 August 2010.
http://www.ihk-siegen.de/IHK-Veranstaltung_in_Duisburg.476.0.html – 27 July 2004.
http://www.clubofrome.org – 06 December 2004.
http://www.oecd.org/std/ tradhome.htm – 15 May 2003.
http://www.basel.int – 09 December 2004.
http://www.polsoz.fu-berlin.de/polwiss/forschung/systeme/ffu/projekte/abgeschlossene/index.html – 13 May 2010.
http://www.britannica.com/eb/article?eu=115396&tocid=10381#10381.toc – 06 September 2010.
http://www.un.org/millenniumgoals/environ.shtml – 03 April 2010.
http://www.unep.or.jp/ietc/index.asp – 03 April 2010.
http://www.utt-gmbh.de –15 July 2010.
http://www.gruenenwald-ag.ch – 15 July 2010.
http://www.sternad.com – 15 July 2010.
http://www.jessberger.de – 15 July 2010.
http://www.ecos-consult.com – 15 July 2010.
http://www.envicom.com – 15 July 2010.
http://www.eco-web.com – 15 July 2010.
http://www.etcentre.org – 15 July 2010.
http://www.cleaner-production.de/en/ – 03 April 2010.
http://archive.greenpeace.org/comms/nukes/chernob/rep02.html – 17 July 2010.
http://ci.nii.ac.jp/ – 15 July 2010.
http://www.unep.org/GC/GC23/documents/GC23-6-add-1.pdf – 15 June 2010.
http://en.wikipedia.org/wiki/Phronesis – 17 July 2010.
http://www.igc.apc.org/habitat/agenda21 – 06 June 2010.
http://www.johannesburgsummit.org/html/documents/summit_docs/2309_planfinal.htm – 17 July 2010.
http://fletcher.tufts.edu/multi/chrono.html – 10 June 2010.
http://sedac.ciesin.org/entri/summaries-menu.html – 10 June 2010.
http://www.au.emb-japan.go.jp/pdf/Whaling.pdf – 02 July 2010.
http://www.icrwhale.org/eng-index.htm – 02 July 2010.

http://www.mofa.go.jp/policy/economy/fishery/index.html – 02 July 2010.
http://www.jfa.maff.go.jp/e/whale/index.html – 02 July 2010.
http://www.guardian.co.uk/environment/2010/apr/08/tokyo-carbon-trading-scheme – 21 July 2010.
http://www.mofa.go.jp/policy/economy/apec/1995/member/info/5.html – 21 September 2010
http://www.earthtrust.org/dnpaper/history.html – 21 September 2010
http://europa.eu.int/scadplus/leg/en/cig/g4000s.htm – 10 July 2010.
http://www.straubing.baynet.de/~k.czauderna/BRD.htm – 10 June 2010.
http://www.route-industriekultur.de/sonstiges/daten-und-fakten/facetten-der-region/der-blaue-himmel ueber-der-ruhr.html – 23 July 2010.
http://www.un.org/millenniumgoals/bkgd.shtml – 02 August 2010.
http://www.mofa.go.jp/policy/oda/white/2009/html/index_shiryo.html – 05 August 2010.
http://www.un.org/documents/ga/conf151/aconf15126-3annex3.htm – 10 July 2010.
http://www.thegef.org – 06 September 2010.
http://www.isda.or.jp/kansai/k1_e.html – 10 September 2010.
http://www.misereor.org/about-us.html – 04 October 2010.
http://www.nabu.de/en/nabu/ – 04 October 2010.
http://www.inwent.org – 5 October 2010

17.3 Press & Radio

Der Spiegel, 1976 – No. 40, 1976.
Der Spiegel, 2010 – 21 June 2010.
Die Zeit, 2010 – 23 June 2010.
Die Zeit, 1998 – 08 January 1998.
Japan Times, 2010 – 22 April 2010.
Japan Times, 1999 – 29 April 1999.
Japan Times, 1995 – 07 January 1995.
NHK 9:00 o'clock news, 1999 – 12 May 1999.
n-tv, 2002 – 21 May 2002.
Sueddeutsche Zeitung, 2002 – 21 May 2002.
Sueddeutsche Zeitung, 2010 – 19 June 2010
Tageszeitung, 2002 – 21 May 2002.
The Economist, 2010 – 01 July 2010.

17.4 Experts

Aichi, Kazuo: Member House of Representatives, Former State Minister for Environment, Tokyo/Japan (02.06.1999)
Breier, Dr. Horst: Head of Evaluation, German Federal Ministry of Economic Cooperation and Development, Bonn/Germany. (18.02.1999)
Franz, Peter: German Federal Ministry for the Environment, Nature Conservation and Nuclear Safety, Berlin/Germany (04.10.1999)
Fuwa, Keiichiro: Professor and Senior Advisor to the Rector, United Nations University, Tokyo/Japan (02.09.1999)
Hashimoto, Dr. Michio: Adviser and former President Overseas Environmental Cooperation Center (OECC), Shiba/Japan. (30.05.1999 & 02.06.1999)
Hiraishi, Taka: Senior Consultant, Institute for Global Environmental Strategies, Hayama/Japan (10.05.1999)
Ishi, Dr. Hiromitsu: Professor and President of Hitotsubashi University, Tokyo/Japan (06.05.1999).
Jensen, Michael: Environmental Attaché, Danish Co-operation for Environment and Development (DANCED), Royal Danish Embassy, Bangkok/Thailand. (30.01.1998)
Kato, Saburo: President, Institute for Environment and Society, Kawasaki/Japan (07.05.1999)
Kawashima, Dr. Yaskuo: Researcher, National Institute for Environmental Studies, Tsukuba/Japan (18.05.1999).
Kaya, Dr. Yoichi: Professor Keio University, General Director Research Institute for New Technologies for the Earth (RITE), Kyoto/Japan (07.05.1999)
Kayama, Yutaka: Associate Director, Research Studies Division, International Centre for Environmental Technology Transfer (ICETT), Yokkaichi/Japan. (15.07.1995)
Kido, Azuma: Director, Global Information Division, Kitakyushu International Techno-Cooperative Association (KITA), Kitakyushu-City/Japan. (10.06.1998)
Kinoshita, Toshio: Director, Environment, WID and other Global Issues Division, Planning Department, JICA, Tokyo/Japan. (15.06.1995 & 17.08.1995)
Kitawaki, Dr. Hidetoshi: Associate Professor, International Planning Centre. Department of Urban Engineering. The University of Tokyo, Tokyo/Japan. (11.07.1995 & 30.06.1998)

Kitterer, Dr. Bernd H.-J.: Managing Director, International Transfer Centre for Environmental Technique (ITUT) & Trade Development and Management Consulting Ltd., Leipzig & Bonn/Germany. (19.12.1998)
Kobori, Dr. Iwao: Professor and Consultant (Environment and Sustainable Development), United Nations University (UNU), Tokyo/Japan. (28.05.1995 & 17.05.1998 & 15.04.1999)
Kulke, Wilhelm: German Environment Foundation (DBU), Special & Executive Director ITUT e.V., Leipzig/Germany (02. June 2010)
Landmann, Ute: Planning Specialist, pilot-project „Strengthening Environmental Technological Capability in Developing Countries (ETC)", German Technical Cooperation (GTZ), Eschborn/Germany. (07.02.1998 & 17.09.1998)
Matsumoto, Hiroshi: Director, Association for Promotion of International Co-operation (APIC), Tokyo/Japan. (05.06.1995)
Matsumoto, Yasuko: Associate Professor, Science University of Tokyo, Suwa College, Chino/Japan. (29.04.1999)
Meyer, Dr. Rolf: Technology Assessment at the German Parliament (TAB), Bonn/Germany. (19.12.1998)
Mitsuhashi, Tadahiro: Senior Chief Editorial Writer, Nihon Keizai Shimbun, Tokyo/Japan (19.05.1999)
Morishima, Akira: Director, International Co-operation Division, Japan Environment Corporation (JEC), Tokyo/Japan. (10.07.1995)
Nakai, Tsuyoshi: Director, Economic Cooperation Policy, Ministry of International Trade and Industry (MITI), Tokyo/Japan. (01.07.1998)
Ohno, Izumi: Director, Environment & Social Development Office, The Overseas Economic Cooperation Fund (OECF), Tokyo/Japan. (03.06.1998 & 10.05.1999)
Ohta, Masahiro: JICA Chief Adviser, Environmental Management Centre (EMC) in collaboration with BAPEDAL and JICA, Serpong/ Indonesia. (15.06.1995 & 23.06.1998)
Onishi, Dr. Akira: Vice President and Professor, Soka University, Tokyo/Japan (09.04.1999)
Pohle, Dr. Horst: Director and Professor, Environmental Technology & Technology Transfer, German Federal Environmental Agency (UBA), Berlin/Germany.(09.12.1998)
Schneider, Alois: German Federal Ministry for Economic Cooperation and Development (BMZ), Bonn/Germany. (19.12.1998)
Schulz, Helmut: German Federal Ministry of Education, Science and Technology (BMBF), Bonn/Germany (18.02.1999)
Suzuki, Tadanori: Director, Environment, WID and other Global Issues Division, Planning Department, JICA, Tokyo/Japan. (21.06.1998)

Tabucanon, Dr. Monthip Sriratana: Director, Environment Research and Training Center (ERTC), Pathumthani/Thailand.(02.02.1998)

Taeger, Uwe: German Federal Ministry for the Environment, Nature Conservation and Nuclear Safety (BMU), Division of Environment & Technique, Technology Transfer, Berlin/Germany (04.10.1999)

Takahashi, Dr. Kazuo: Director, International Development Research Institute (IDRI), Foundation for Advanced Studies on International Development (FASID), Tokyo/Japan. (05.07.1998)

Tanikawa, Kiyoshi: Research and Planning Division, Economic Cooperation Bureau, Ministry of Foreign Affairs, Tokyo/Japan. (18.06.1998)

Tsuji, Masami: Deputy Director, Office of Overseas Environment Cooperation, Global Environment Department, Environment Agency (EA), Tokyo/Japan. (17.06.1998)

Uemachi, Tohuru: First Technical Cooperation Division, Social Development Cooperation Department, JICA, Tokyo/Japan. (30.06.1998)

Ueta, Dr. Kazuhiro: Professor of Environmental Economics and Public Finance, Graduate School of Economics, Kyoto University, Kyoto/Japan. (13.04.1999)

Ui, Jun: Professor of Environmental Sciences and Sanitary Engineering University of Okinawa, Naha/Japan. (01.05.1999 & 22.05.1999)

Usuki, Mitsuo: Senior Advisor to Director General, Global Environment Department, Japan Environment Agency, Tokyo/Japan (22.10.1999)

Yasui, Dr. Itaru: Professor, Institute of Industrial Sciences (IIS) and Director, Center for Collaborative Research, Tokyo University, Tokyo/Japan. (03.07.1998)

18 Appendix

18.1 Index of Japanese terms

Japanese Terms	English Translation
chiri mo tsumoreba yama to naru	Every piece of dust one day becomes a mountain; Many a little makes a mickle
gankō keitai hattenron	Flying Geese Pattern of Development
jizokuteki kaihatsu or *jizoku kano na kaihatsu*	Sustainable development or economic development
jizokuteki hatten	Sustainable development or economic development (same as above, shortened version)
kankyō	Environment
kōgai	Pollution (environment)
kankyōchō	Japan Environmental Agency
kōgai boshichō	Pollution Prevention Agency
kagaku gijutsu	Science & Technology
tekunorojī	Technology
kankyō gijutsu	Environmental Technology (ET)
bakufu	Shogunate
shōgun	Japanese military rank and historical title; literally, "a commander of a force"
zaibatsu	Family groups heading banks, industry and trade
harakiri	Japanese ritual suicide
kokusai kokka	International State
joseikin	Grants
rōn	Loans
Kokusai kyōryoku ginkō	Japan Bank for International Cooperation
kakugi kettei	Cabinet Decision
kakugi kettei – Ajia shokoku ni tai suru keizai kyōryoku ni kan suru ken	Policy on Economic Cooperation with the Countries of Asia
Ajia keizai kondankai	Asian Economic Deliberation Council
chihō seifu	Local Administrations in Japan
chūō seifu	Central Government
konsarutanto gaisha	Consulting Agency
hi seifu soshiki	NGO

18.2 Index of German terms

German Terms	English translation
Umwelt	Environment
Umweltverschmutzung	Pollution (environment)
Technologie	Technology
Umwelttechnologie	Environmental Technology (ET)
Internationales Transferzentrum für Umwelttechnik	Centre for International Transfer of Environmental Techniques
Grundgesetz	Constitution
Bund für Vogelschutz (BfV)	German Society for Bird Protection
Reichstag	Diet
Grundlagenvertrag	Basis-of-Relations-Agreement
Bundes-Immissionsschutzgesetzes	German Law on Pollutants
Wasserhaushaltsgesetz	Water Management Law
Verfassungsnorm	Constituency Norm
Landeskulturgesetz	National Culture Law
Sozialistische Einheitspartei Deutschlands	Socialist Unity Party
Umweltbundesamt	Environmental Agency
Großfeuerungsanlagenverordnung	Combustion Plants
Bundesministerium für wirtschaftliche Zusammenarbeit (BMZ)	Federal Ministry of Economic Cooperation
Finanzielle Zusammenarbeit	Financial Cooperation
Technische Zusammenarbeit	Technical Cooperation
Kreditanstalt für Wiederaufbau	Development Loan Cooperation
Deutsche Gesellschaft für Technische Zusammenarbeit	German Agency for Technical Cooperation
Außenhandelskammern	German Chamber Network

18.3 Japan - International environmental treaties in force

Date Entered into Force	Date of Signature	Title
27.09.1980	28.12.1972	Convention for the Conservation of Antarctic Seals
21.04.1951		International Convention for the Regulation of Whaling
01.11.1982		Agreed Measures for the Conservation of Antarctic Fauna and Flora
01.07.1970		Convention for the Establishment of an Inter American Tropical Tuna Commission
21.03.1969	28.10.1966	International Convention for the Conservation of Atlantic Tunas
15.02.1979		Protocol amending the International Convention for the High Seas Fisheries of the North Pacific Ocean
15.07.1977		Convention on the International Regulations for Preventing Collisions at Sea
06.05.1975	15.12.1970	International Convention relating to Intervention on the High Seas in Cases of Oil Pollution Casualties
02.10.1983		International Convention for the Prevention of Pollution from Ships as modified by the Protocol of 1978
16.10.1978		International Convention on the Establishment of an International Fund for Compensation for Oil Pollution Damage
25.05.1980		International Convention for the Safety of Life at Sea (SOLAS)
01.12.1986		Convention on Limitation of Liability for Maritime Claims
22.11.1994		Protocol to the International Convention on Civil Liability for Oil Pollution Damage
01.09.1976		International Convention on Civil Liability for Oil Pollution Damage
11.0.1952	06.12.1951	International Plant Protection Convention
07.04.1982	12.09.1980	Convention on the Conservation of Antarctic Marine Living Resources
21.11.1967		International Convention for the Prevention of Pollution of the Sea by Oil,1954, as amended in 1962 and 1969
09.06.1982		Convention on the Prohibition of Military or any other Hostile Use of Environmental Modification Techniques
10.07.1968		Convention on the High Seas
20.06.1983		Convention on Registration of Objects Launched into Outer Space
27.01.1930		International Agreement for the Creation of an International Office for dealing with Contagious Diseases of Animals at Paris

10.10.1967	27.01.1967	Treaty on Principles Governing the Activities of States in the Exploration and Use of Outer Space, including the Moon and other Celestial Bodies
18.05.1972	11.02.1971	Treaty on the Prohibition of the Emplacement of Nuclear Weapons and other Weapons of Mass Destruction on the Sea Bed and the Ocean Floor and in the Subsoil thereof
20.06.1983		Convention on International Liability for Damage caused by Space Objects
15.06.1964	14.08.1963	Treaty Banning Nuclear Weapon Tests in the Atmosphere, in Outer Space and under Water
08.06.1982	10.04.1972	Convention on the Prohibition of the Development, Production and Stockpiling of Bacteriological (Biological) and Toxin Weapons and on their Destruction
14.11.1980	22.06.1973	Convention on the Prevention of Marine Pollution by Dumping of Wastes and Other Matter
26.07.1978		Convention concerning Prevention and Control of Occupational Hazards caused by Carcinogenic Substances and Agents (ILO No. 139)
31.07.1974		Convention concerning the Protection of Workers against Ionising Radiations (ILO No. 115)
23.06.1961	01.12.1959	The Antarctic Treaty
29.07.1957		Statute of the International Atomic Energy Agency
08.10.1953		Convention on International Civil Aviation Annex 16 Aircraft Noise
04.05.1959	19.11.1956	Protocol to the International Convention for the Regulation of Whaling
10.09.1953		Convention of the World Meteorological Organization
04.11.1980	30.04.1973	Convention on International Trade in Endangered Species of Wild Fauna and Flora
17.10.1980		Convention on Wetlands of International Importance especially as Waterfowl Habitat
12.06.1979		International Convention for Safe Container (CSS)
20.09.1996	07.02.1983	United Nations Convention on the Law of the Sea
14.11.1980		Amendments to Annexes to the Convention on the Prevention of Marine Pollution by Dumping of Wastes and Other Matter concerning Incineration at Sea
		Amendments to the Convention on the Prevention of Marine Pollution by Dumping of Wastes and Other Matter concerning Settlement of Disputes
26.06.1987		Protocol to amend the Convention on Wetlands of International Importance especially as Waterfowl Habitat
01.04.1985	28.03.1984	International Tropical Timber Agreement
02.09.1984	03.08.1984	Provisional Understanding Regarding Deep Seabed Matters

28.12.1988		Convention for the Protection of the Ozone Layer
	24.04.1987	Convention on the Law of Treaties between States and International Organizations or between International Organizations
17.12.1984	17.12.1984	Protocol relating to modification of the International Convention for the Conservation of Atlantic Tunas (Agreement on the Protection of Confidentiality of Data related to Deep Sea bed Areas for which application of Authorisation has been made)
10.07.1987	06.03.1987	Convention on Early Notification of a Nuclear Accident
10.07.1987	06.03.1987	Convention on Assistance in the Case of a Nuclear Accident or Radiological Emergency
01.01.1989	16.09.1987	Protocol on Substances that deplete the Ozone Layer
	22.11.1989	Convention on the Regulation of Antarctic Mineral Resource Activities
01.07.1992		International Convention for the Prevention of Pollution from Ships, 1973 (MARPOL) Annex III (Optional): Hazardous substances carried in packaged form
		International Convention for the Prevention of Pollution from Ships (MARPOL) Annex IV (Optional): Sewage
31.12.1988		International Convention for the Prevention of Pollution from Ships (MARPOL) Annex V (Optional) = Garbage
17.12.1993		Convention on the Control of Transboundary Movements of Hazardous Wastes and their Disposal
13.04.1987		Amendment to the Convention on International Trade in Endangered Species of Wild Fauna and Flora (art.XI)
01.05.1994		Amendments to Articles 6 and 7 of the Convention on Wetlands of International Importance especially as Waterfowl Habitat
10.08.1992		Amendment to the Montreal Protocol on Substances that deplete the Ozone Layer
24.03.1992	20.12.1991	Convention establishing a marine scientific organization for the North Pacific Region (PICES)
17.01.1996		International Convention on Oil Pollution Preparedness, Response and Co operation
	29.09.1992	Protocol to the Antarctic Treaty on Environmental Protection
21.03.1994	13.06.1992	Framework Convention on Climate Change
29.12.1993	13.06.1992	Convention on Biological Diversity
20.03.1995		Amendment to the Montreal Protocol on Substances that deplete the Ozone Layer
30.05.1996		Protocol to amend the International Convention on the Establishment of an International Fund for Compensation for Oil Pollution Damage

30.05.1996		Protocol to amend the International Convention on Civil Liability for Oil Pollution Damage
03.09.1982	1710.1979	International Convention for the Protection of New Varieties of Plants as amended on 23.10.1978
01.01.1997	13.12.1994	International Tropical Timber Agreement
28.07.1996	29.07.1994	Agreement relating to the Implementation of Part XI of the United Nations Convention on the Law of the Sea of 10 December 1982
26.07.1996	14.10.1994	International Convention to combat Desertification in those Countries Experiencing Serious Drought and/or Desertification, particularly in Africa
	20.09.1994	Convention on Nuclear Safety
26.06.1996		Agreement for the Establishment of the Indian Ocean Tuna Commission

(Composed out of data of ENTRI 2010)

18.4 Germany - International environmental treaties in force

Date Entered into Force	Date of Signature	Title
30.10.1987		Convention for the Conservation of Antarctic Seals
06.12.1905		Convention for the Protection of Birds Useful to Agriculture
07.06.1886		Treaty concerning the Regulation of Salmon Fishery in the Rhine River Basin
02.07.1982		International Convention for the Regulation of Whaling
01.03.1966	20.12.1962	Agreement on the Protection of the Salmon in the Baltic Sea
05.08.1975	29.11.1969	International Convention relating to Intervention on the High Seas in Cases of Oil Pollution Casualties
02.10.1983	16.11.1978	International Convention for the Prevention of Pollution from Ships as modified by the Protocol of 1978
	04.03.1974	International Convention for the Prevention of Pollution from Ships (MARPOL)
19.11.1985	04.03.1974	Protocol relating to Intervention on the High Seas in Cases of Pollution by Substances other than Oil
30.12.1975	17.12.1971	Convention Relating to Civil Liability in the Field of Maritime Carriage of Nuclear Material
08.04.1981	16.12.1977	Protocol to the International Convention on Civil Liability for Oil Pollution Damage
18.08.1975	29.11.1969	International Convention on Civil Liability for Oil Pollution Damage
26.03.1973		Constitution of the European Commission for the Control of Foot and Mouth Disease
03.05.1957	30.04.1952	International Plant Protection Convention
24.11.1976	21.01.1972	Protocol amending the Agreement on the Protection of the Salmon in the Baltic Sea
23.05.1982	11.09.1980	Convention on the Conservation of Antarctic Marine Living Resources
26.07.1958		International Convention for the Prevention of Pollution of the Sea by Oil,1954, as amended in 1962 and 1969
25.08.1973	30.10.1958	Convention on the High Seas
20.08.1987	30.04.1970	Vienna Convention on the Law of Treaties
16.10.1979	02.03.1976	Convention on Registration of Objects Launched into Outer Space
	30.10.1958	Convention on the Continental Shelf
01.01.1970	13.12.1957	European Agreement concerning the International Carriage of Dangerous Goods by Road (ADR)

19.04.1985		Protocol amending the European Agreement concerning the International Carriage of Dangerous Goods by Road (ADR)
16.02.1928		International Agreement for the Creation of an International Office for dealing with Contagious Diseases of Animals at Paris
03.05.1980	22.03.1974	Convention on the Protection of the Marine Environment of the Baltic Sea Area
06.10.1991	13.06.1980	Convention on the Physical Protection of Nuclear Material
10.02.1971	27.01.1967	Treaty on Principles Governing the Activities of States in the Exploration and Use of Outer Space, including the Moon and other Celestial Bodies
07.04.1983	10.04.1972	Convention on the Prohibition of the Development, Production and Stockpiling of Bacteriological (Biological) and Toxin Weapons and on their Destruction
08.12.1977	22.01.1973	Convention on the Prevention of Marine Pollution by Dumping of Wastes and Other Matter
30.01.1959		Amendment of the Convention for the Regulation of the Meshes of Fishing Nets and the Size Limits of Fish (Article 7, paragraph 2)
06.05,1960		Amendments to the Convention for the Regulation of the Meshes of Fishing Nets and the Size Limits of Fish
09.05.1961	09.05.1961	Amendments to the Convention for the Regulation of the Meshes of Fishing Nets and the Size Limits of Fish
11.05.1962		Amendments to the Convention for the Regulation of the Meshes of Fishing Nets and the Size Limits of Fish
01.06.1963		Amendments to the Convention for the Regulation of the Meshes of Fishing Nets and the Size Limits of Fish
26.09.1976	15.11.1967	Convention on Conduct of Fishing Operations in the North Atlantic
27.06.1963	24.01.1959	North East Atlantic Fisheries Convention
19.01.1970	09.03.1964	Fisheries Convention
	28.04.1978	Convention on Civil Liability for Oil Pollution Damage resulting from Exploration for and Exploitation of Seabed Mineral Resources
30.09.1975	29.07.1960	Convention on Third Party Liability in the Field of Nuclear Energy
22.07.1959		Convention on the Establishment of a Security Control in the Field of Nuclear Energy
30.09.1975	28.01.1964	Additional Protocol to the Convention on Third Party Liability in the Field of Nuclear Energy
18.11.1994		Convention concerning the Protection of Workers against Occupational Hazards in the Working Environment due

to Air Pollution, Noise and Vibration (ILO No. 148)

28.11.1981		Convention Concerning Minimum Standards in Merchant Ships (No. 147)
23.08.1977		Convention concerning Prevention and Control of Occupational Hazards caused by Carcinogenic Substances and Agents (ILO No. 139)
26.09.1974		Convention concerning Protection against Hazards of Poisoning arising from Benzene (ILO No.136)
26.09.1974		Convention concerning the Protection of Workers against Ionising Radiations (ILO No. 115)
02.03.1973	16.09.1968	European Agreement on the Restriction of the Use of certain Detergents in Washing and Cleaning Products
01.10.1957		Statute of the International Atomic Energy Agency
08.06.1956		Convention on International Civil Aviation Annex 16 Aircraft Noise
10.11.1961		Agreement on the Protection of Lake Constance against Pollution
31.12.1956		Convention on the Canalization of the Mosel
01.07.1962		Protocol concerning the Constitution of an International Commission for the Protection of the Mosel against Pollution
02.07.1977		Protocol amending the Convention on the Canalization of the Mosel
01.07.1974	13.12.1968	European Convention for the Protection of Animals during International Transport
07.11.1989	10.05.1979	Additional Protocol to the European Convention for the Protection of Animals during International Transport
02.07.1982		Protocol to the International Convention for the Regulation of Whaling
10.07.1954		Convention of the World Meteorological Organization
01.04.1985	19.09.1979	Convention on the Conservation of European Wildlife and Natural Habitats
12.11.1975	13.08.1970	Protocol to the Convention for the International Council for the Exploration of the Sea
20.06.1976	03.03.1973	Convention on International Trade in Endangered Species of Wild Fauna and Flora
01.05.1965	29.04.1963	Agreement concerning the International Commission for the Protection of the Rhine against Pollution
01.02.1979	03.12.1976	Supplementary Agreement to the 1963 Agreement on the International Commission for the Protection of the Rhine against Pollution
25.11.1967	30.04.1966	Convention regulating the Withdrawal of Water from Lake Constance
01.02.1979	03.12.1976	Convention for the Protection of the Rhine against

		Chemical Pollution
05.07.1985	03.12.1976	Convention for the Protection of the Rhine from Pollution by Chlorides modified by Exchanges of letters
26.06.1954		Convention for the Establishment of the European and Mediterranean Plant Protection Organisation
01.10.1984	23.06.1979	Convention on the Conservation of Migratory Species of Wild Animals
07.01.1959		Convention on the International Maritime Organization
16.03.1983	13.11.1979	Convention on Long Range Transboundary Air Pollution
26.06.1976	28.11.1974	Convention on Wetlands of International Importance especially as Waterfowl Habitat
09.10.1977		Convention on Fishing and Conservation of the Living Resources in the Baltic Sea and the Belts
01.01.1979	24.11.1978	Convention on Future Multilateral Cooperation in the Northwest Atlantic Fisheries (NAFO)
22.11.1994	19.11.1976	Protocol to the International Convention on the Establishment of an International Fund of Compensation for Oil Pollution Damage
16.09.1972		Regulation No. 15 : Uniform Provisions concerning Approval of Vehicles Equipment with a Positive Ignition Engine with regard to Emission of Gaseous Pollutants by the Engine
13.11.1973		Regulation No. 24 : Uniform Provisions concerning the Approval of Vehicles Equipped with Diesel Engines with regard to the Emission of Pollutants by the Engine
25.10.1975		Regulation No. 28 : Uniform Provisions for the Approval of Audible Warning Devices and of Motor Vehicles with regard to their Audible Signals
13.06.1983		Regulation No. 40 : Uniform Provisions concerning Approval of Motor Cycles Equipped with a Positive Ignition Engine with regard to the Emission of Gaseous Pollutants by the Engine
03.10.1990		Regulation No. 41 : Uniform Provisions concerning the Approval of Motor Cycles with regard to Noise
19.01.1976		Agreement on an International Energy Program Amendments to the International Convention for the Prevention of Pollution of the Sea by Oil,1954,concerning the Protection of the Great Barrier Reef
11.06.1954		Convention for the Regulation of the Meshes of Fishing Nets and the Size Limits of Fish
01.11.1981		Regulation No. 47 : Uniform Provisions concerning the Approval of Mopeds equipped with a Positive Ignition Engine with regard to the Emission of Gaseous Pollutants by the Engine

18 Appendix

02.09.1982	02.09.1982	Agreement concerning Interim Arrangements relating to Polymetallic Nodules of the Deep Sea Bed
		Amendments to the Convention on the Prevention of Marine Pollution by Dumping of Wastes and Other Matter concerning Settlement of Disputes
01.10.1986	13.01.1983	Protocol to amend the Convention on Wetlands of International Importance especially as Waterfowl Habitat
	25.10.1983	Protocol amending the European Agreement on the Restriction of the Use of certain Detergents in Washing and Cleaning Products
01.06.1985	21.06.1983	Second Protocol amending the Convention on the Canalization of the Mosel
10.02.1984		Amendments to the Convention on Fishing and Conservation of the Living Resources in the Baltic Sea and the Belts
21.03.1986	29.06.1984	International Tropical Timber Agreement
15.12.1985		Regulation No. 49 :Uniforms Provisions concerning the Approval of Diesel Engines with regard to the Emission of Gaseous Pollutants
03.10.1990		Regulation No. 51: Uniform provisions concerning the approval of motor vehicles having at least four wheels with regard to their noise
01.09.1989	13.09.1983	Agreement for Cooperation in Dealing with Pollution of the North Sea by Oil and other Harmful Substances
01.09.1989	02.03.1983	Protocol amending the Convention for the Prevention of Marine Pollution by Dumping from Ships and Aircraft
02.09.1984	03.08.1984	Provisional Understanding Regarding Deep Seabed Matters
28.01.1988	26.02.1985	Protocol to the Convention on Long range Transboundary Air Pollution on Long Term Financing of Co operative Programme for Monitoring and Evaluation of the Long Range Transmission of Air Pollutants in Europe (EMEP)
29.12.1988	22.03.1985	Convention for the Protection of the Ozone Layer
	25.05.1984	Protocol to amend the International Convention on Civil Liability for Oil Pollution Damage
15.10.1989	26.09.1986	Convention on Early Notification of a Nuclear Accident
15.10.1989	26.09.1986	Convention on Assistance in the Case of a Nuclear Accident or Radiological Emergency
01.09.1989	26.03.1986	Protocol amending the Convention for the prevention of marine pollution from land based sources
01.01.1989	16.11.1987	Protocol on Substances that deplete the Ozone Layer
01.01.1988	12.05.1987	Third Protocol amending the Convention on the canalization of the Mosel
01.07.1992		International Convention for the Prevention of Pollution from Ships, 1973 (MARPOL) Annex III (Optional):

		Hazardous substances carried in packaged form
		International Convention for the Prevention of Pollution from Ships (MARPOL) Annex IV (Optional): Sewage
31.12.1988		International Convention for the Prevention of Pollution from Ships (MARPOL) Annex V (Optional) = Garbage
14.02.1991	01.11.1988	Protocol to the Convention on Long Range Transboundary Air Pollution concerning the Control of Emissions of Nitrogen Oxides or their Transboundary Fluxes
20.07.1995	23.10.1989	Convention on the Control of Transboundary Movements of Hazardous Wastes and their Disposal
13.04.1987		Amendment to the Convention on International Trade in Endangered Species of Wild Fauna and Flora (art.XI)
01.05.1992	21.06.1988	European Convention for the Protection of Pet Animals
01.11.1991	21.06.1988	European Convention for the Protection of Vertebrate Animals used for Experimental and other Scientific Purposes
11.05.1990	13.06.1985	Amendments to Annexes I and II to the Convention for the Prevention of Marine Pollution by Dumping from Ships and Aircraft
	05.12.1989	Protocol amending the Convention for the Prevention of Marine Pollution by Dumping from Ships and Aircraft Amendment to the Convention on International Trade in Endangered Species of Wild Fauna and Flora (Art.XXI)
01.05.1994		Amendments to Articles 6 and 7 of the Convention on Wetlands of International Importance especially as Waterfowl Habitat
01.10.1991	13.10.1990	Agreement on the Conservation of Seals in the Wadden Sea
10.08.1992		Amendment to the Montreal Protocol on Substances that deplete the Ozone Layer
05.11.1989		Regulation No. 83 :Uniforms Provisions concerning the Approval of Vehicles with regard to the Emission of Gaseous Pollutants by the Engine according to the Engine Fuel Requirements
	26.02.1991	Convention on Environmental Impact Assessment in a Transboundary Context
15.05.1995	30.11.1990	International Convention on Oil Pollution Preparedness, Response and Co operation
	08.10.1990	Convention on the international commission for the protection of the Elbe
10.08.1968	02.12.1961	International Convention for the Protection of New Varieties of plants
	19.03.1991	International Convention for the Protection of New Varieties of Plants (consolidated version)

18 Appendix

	04.10.1991	Protocol to the Antarctic Treaty on Environmental Protection
04.04.1991		International Plant Protection Convention (Revised Text)
06.03.1995	07.11.1991	Convention concerning the Protection of Alps
	19.11.1991	Protocol to the 1979 Convention on Long range Transboundary Air Pollution concerning the Control of Emissions of Volatile Organic Compounds or their Transboundary Fluxes
29.03.1994	09.04.1992	Agreement on the Conservation of Small Cetaceans of the Baltic and North Sea
06.10.1996	18.03.1992	Convention on the Protection and Use of Transboundary Watercourses and International Lakes
	18.03.1992	Convention on Transboundary Effects of Industrial Accidents
21.03.1994	12.06.1992	Framework Convention on Climate Change
21.03.1994	12.06.1992	Convention on Biological Diversity
	22.09.1992	Convention for the Protection of the Marine Environment of the North East Atlantic
	09.04.1992	Convention on the Protection of the Marine Environment of the Baltic Sea Area
16.01.1994	05.12.1991	Agreement on the Conservation of Bats in Europe
	13.01.1993	Convention on the Prohibition of the Development, Production, Stockpiling and Use of Chemical Weapons and their Destruction
14.06.1994		Amendment to the Montreal Protocol on Substances that deplete the Ozone Layer
18.11.1994		Convention concerning Safety in the Use of Asbestos
	30.06.1994	Protocol to the International Convention for the Safety of Fishing Vessels
12.01.1992		Regulation No. 84 :Uniforms Provisions concerning the Approval of Passengers Cars Equipped with an Internal Combustion Engine with regard to the Measurement of Fuel Consumption
30.05.1996		Protocol to amend the International Convention on the Establishment of an International Fund for Compensation for Oil Pollution Damage
30.05.1996		Protocol to amend the International Convention on Civil Liability for Oil Pollution Damage
01.11.1994	25.09.1991	Protocol additional to the Convention for the Protection of the Rhine from Pollution by Chlorides
12.11.1986	23.10.1978	International Convention for the Protection of New Varieties of Plants as amended on 23.10.1978
	14.06.1994	Protocol to the Convention on Long Range Transboundary Air Pollution on further Reduction of Sulphur Emissions
30.08.1995	30.08.1995	International Tropical Timber Agreement

26.12.1996	14.10.1994	International Convention to combat Desertification in those Countries Experiencing Serious Drought and/or Desertification, particularly in Africa
	20.09.1994	Convention on Nuclear Safety
	29.06.1994	Convention on Cooperation for the Protection and Sustainable Use of the Danube River
	17.12.1994	Energy Charter Treaty
	20.12.1994	Protocol for the implementation of the Alpine Convention in the field of town and country planning and sustainable development
	20.12.1994	Protocol for the implementation of the Alpine Convention in the field of mountain agriculture
	20.12.1994	Protocol for the implementation of the Alpine Convention in the field of nature protection and landscape conservation
01.04.1994		Amendments of the Agreement for Co operation in dealing with Pollution by Oil and other Harmful Substances
	15.08.1996	Agreement on the Conservation of African Eurasian Migratory Waterbirds
	17.12.1994	Energy Charter Protocol on Energy Efficiency and related Environmental Aspects

(Composed out of data of ENTRI 2010)

Arbeit, Bildung & Gesellschaft
Labour, Education & Society

Herausgegeben von Prof. Dr. György Széll, Prof. Dr. Heinz Sünker,
Dr. Anne Inga Hilsen und Dr. Francesco Garibaldo

Bd. 1   György Széll (ed.): Corporate Social Responsibility in the EU & Japan. 2006.

Bd. 2   Katja Maar: Zum Nutzen und Nichtnutzen der Sozialen Arbeit am exemplarischen Feld der Wohnungslosenhilfe. Eine empirische Studie. 2006.

Bd. 3   Daniela De Ridder: Vom urbanen Sozialraum zur kommunikativen Stadtgesellschaft. 2007.

Bd. 4   Heinz Sünker / Ingrid Miethe (Hrsg.): Bildungspolitik und Bildungsforschung: Herausforderungen und Perspektiven für Gesellschaft und Gewerkschaften in Deutschland. 2007.

Bd. 5   Anja Bastigkeit: Bildungsbiographie und elementarpädagogische Bildungsarbeit. 2007.

Bd. 6   Antônio Inácio Andrioli: Biosoja versus Gensoja. Eine Studie über Technik und Familienlandwirtschaft im nordwestlichen Grenzgebiet des Bundeslandes Rio Grande do Sul (Brasilien). 2007.

Bd. 7   Russell Farnen / Daniel German / Henk Dekker / Christ'l De Landtsheer / Heinz Suenker (eds.): Political Culture, Socialization, Democracy, and Education. Interdisciplinary and Cross-National Perspectives for a New Century. 2008.

Bd. 8   Francesco Garibaldo / Volker Telljohann (eds.): New Forms of Work Organisation and Industrial Relations in Southern Europe. 2007.

Bd. 9   Anne Marie Berg / Olav Eikeland (eds.): Action Research and Organisation Theory. 2008.

Bd. 10   György Széll / Carl-Heinrich Bösling / Ute Széll (eds.): Education, Labour & Science. Perspectives for the 21st Century. 2008.

Bd. 11   Francesco Garibaldo / Philippe Morvannou / Jochen Tholen (eds.): Is China a Risk or an Opportunity for Europe? An Assessment of the Automobile, Steel and Shipbuilding Sectors. 2008.

Bd. 12   Yunus Dauda: Managing Technology Innovation. The Human Resource Management Perspective. 2009.

Bd. 13   Jarmo Lehtonen / Satu Kalliola (eds.): Dialogue in Working Life Research and Development in Finland. 2009.

Bd. 14   György Széll / Werner Kamppeter / Woosik Moon (eds.): European Social Integration – A Model for East Asia? 2009.

Bd. 15   Benedicte Brøgger / Olav Eikeland (eds.): Turning to Practice with Action Research. 2009.

Bd. 16   Till Johannes Hoffmann: Verschwendung. Philosophie, Soziologie und Ökonomie des Überflusses. 2009.

Bd. 17   Denis Harrisson / György Széll / Reynald Bourque (eds.): Social Innovation, the Social Economy and World Economic Development. 2009.

Bd. 18   Werner Weltgen: Total Quality Management als Strukturierungsaufgabe für nachhaltigen Unternehmenswandel. 2009.

Bd. 19   György Széll / Ute Széll (eds.): Quality of Life and Working Life in Comparison. 2009.

Bd. 20   Francesco Garibaldo / Volker Telljohann (eds.): The Ambivalent Character of Participation. New Tendencies in Worker Participation in Europe. 2010.

Band 21 Richard Ennals / Robert H. Salomon (eds.): Older Workers in a Sustainable Societey. 2011.

Band 22   Christoph Sänger: Anna Siemsen – Bildung und Literatur. 2011.

Band 23   Nam-Kook Kim: Deliberative Multiculturalism in Britain. A Response to Devolution, European Integration, and Multicultural Challenges. 2011.

Band 24   Mirella Baglioni / Bernd Brandl (eds.): Changing Labour Relations. Between Path Dependency and Global Trends. 2011.

Band 25   Rüdiger Kühr: Japan`s Transnational Environmental Policies. The Case of Environmental Technology Transfer to Newly Industrializing Countries. 2011.

www.peterlang.de

www.ingramcontent.com/pod-product-compliance
Ingram Content Group UK Ltd.
Pitfield, Milton Keynes, MK11 3LW, UK
UKHW021822140426
5217IPUK00004B/53